The Deep Sea

BRUCE ROBISON
AND
JUDITH CONNOR

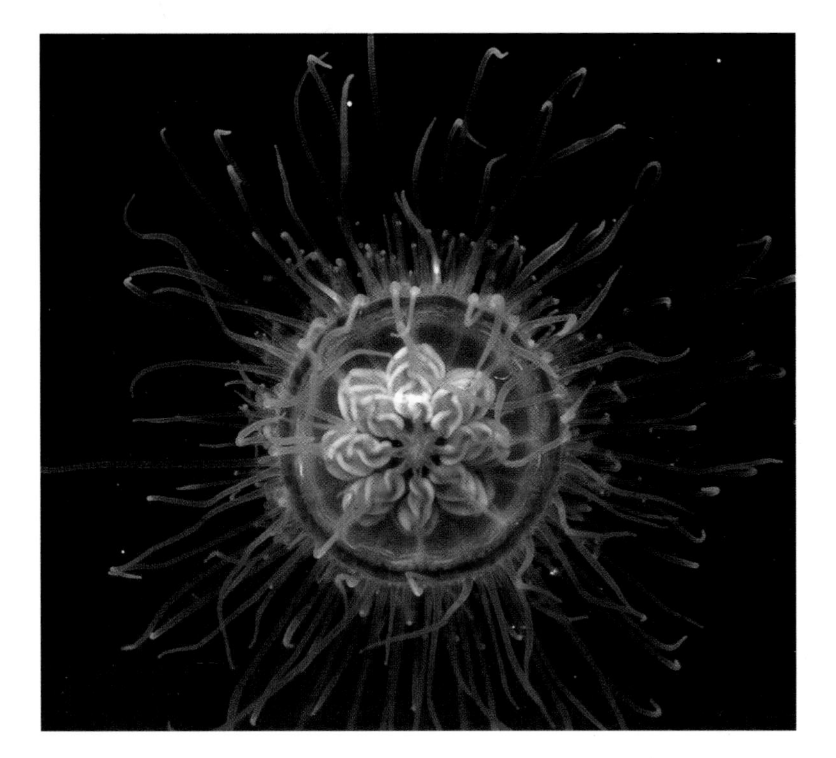

MONTEREY BAY AQUARIUM PRESS
Monterey, California

The mission of the Monterey Bay Aquarium is to inspire conservation of the oceans.

The Monterey Bay Aquarium wishes to acknowledge the tremendous assistance provided by its sister institution, the Monterey Bay Aquarium Research Institute, in interpreting the deep-sea world. (www.mbari.org)

Acknowledgments We thank Kim Reisenbichler, Steve Etchemendy, Chuck Baxter and Annette Gough, for making this book possible. Thanks also to MBARI's Division of Marine Operations, especially the crew of the *R/V Point Lobos* and our pilots, for enabling these deep explorations. Bill Hamner, and Nancy Harray contributed more than they may know.

Dedication We dedicate this book to David Packard with deepest gratitude, and to Alexander Vassar with highest hopes.

Published in the United States by the Monterey Bay Aquarium Foundation, 886 Cannery Row, Monterey, CA 93940-1085 www.mbayaq.org

Library of Congress Cataloging in Publication Data
Robison, Bruce H.
The deep sea / Bruce Robison and Judith Connor.
p. cm. – (Monterey Bay Aquarium natural history series)
Includes index.
ISBN 1-878244-25-6
1. Deep-sea biology. 2. Abyssal zone. 3. Deep-sea biology—California—Monterey Bay
II. Connor, Judith. II. Monterey Bay Aquarium.
III. Title. IV. Series.
QH91.R63 1999
578.779—dc21 98-47557
CIP

ISBN 1-878244-25-6

All photos and illustrations copyright Monterey Bay Aquarium except:

Ancient Art & Architecture Collection Ltd.: 4-5

Barry, James/MBARI: 25 (bottom), 60

Bostelmann, Else/National Geographic Image Coll.: 6 (top), 19 (middle & bottom)

Challenger Reports: 5 (top), 14 (top)

Gowlett-Holmes, Karen G.: 76 (top)

Hessler, Robert/Scripps Institution of Oceanography: 65 (bottom)

JAMSTEC: 25 (top), 73 (top)

Knudsen, David/National Geographic Image Coll.: 16 (bottom)

Lanting, Frans/Minden Pictures: 56

Leet, M., for MBARI: 21 (top left & right)

Lyle, Jeremy/Tasmanian Aquaculture & Fisheries Inst.: 75 (bottom)

Madin, Larry/ Woods Hole Oceanographic Inst.: 33, 52 (top)

Maher, Norman/MBARI, Graphic by: 26

Matsumoto, George I./MBARI: 34 (top), 50 (bottom), 51 (bottom)

Matsumoto, George I.: 10 (top right), 50 (top), 51 (top)

Mills, Claudia: 8 (top)

Monterey Bay Aquarium Research Institute (MBARI): 23, 40 (top), 45, 49 (bottom), 53 (top), 57 (top), 59, 61 (bottom), 71, 72, 79 (top)

Murray, Dawn/MBARI: back cover (left), 77 (top)

NASA: 4

Nybakken, James: 39 (middle), 77 (bottom)

Orange, Dan/MBARI: 75 (top)

Raskoff, Kevin/MBARI: 43 (top), 52 (bottom), 44 (top)

Reisenbichler, Kim/MBARI: 54 (bottom), 55

Richard Ellis Collection, Courtesy of: 5 (bottom right), 47

Richard Ellis Collection/U.S. Navy Submarine Force Museum: 18 (top)

Riddiford, Charles/National Geographic Image Coll.: 16 (top right)

Robison, Bruce H./Sea Studios: 6 (bottom)

Robison, Bruce H.: cover, 9 (top),13, 15, 18 (middle & bottom), 34 (bottom), 38, 39 (top), 42 (bottom), 53 (bottom), 57 (bottom), 70 (top)

SAIC: 10 (bottom)

Sea Studios: 8 (bottom)

Sherlock, Rob/MBARI: 49 (top left)

Smoyer, T./ Harbor Branch Oceanographic Inst.: 40 (bottom right)

Somero, George, Courtesy of: 74

Stakes, Debra: 22 (top),27, 28

Tee-Van, John/National Geographic Image Coll: 19 (top)

Webster, Steven K.: 67 (top)

White, Marv/MBARI: 11

Widder, E./ Harbor Branch Oceanographic Inst.: 40 (bottom left), 53 (top), 78 (top)

Wildlife Conservation Society (Bronx Zoo): 16 (top left), 17

Woods Hole Oceanographic Inst.: 65 (top)

Wrobel, David J./MBARI: 29, 30, 61 (top left), 62, 63, 64 (bottom), 68 (top), 70 (middle), 76 (bottom)

Wrobel, David J.: back cover (right), 1, 34 (middle), 42 (top), 48, 61 (top right), 66 (middle), 69, 73 (bottom)

Wu, Norbert: 37 (bottom)

Youngbluth, M./ Harbor Branch Oceanographic Inst.: 49 (top right), 68 (bottom)

Series and Book Editor: Nora L. Deans
Project Editor: Roxane Buck-Ezcurra
Contributing Writers: Natasha Fraley, Christina J. Slager, Steven K. Webster
Design: Ann W. Douden
Printed on recycled papter in Hong Kong by Global Interprint

Contents

1 **Exploring the Deep Sea** 6

2 **Challenges of Life in the Deep** 22

3 **Life in the Ocean's Midwaters** 34

4 **Life on the Deep Seafloor** 58

5 **People and the Deep** 74

A Sampling of Deep-Sea Species 78

Index 80

The deep sea looms dark and mysterious—fascinating us over the centuries with visions of sea serpents, giant squids, bizarre fishes and frigid, dark waters that guard their secrets. While images from space vividly illustrate how much of Earth is covered by oceans, they tell us little of what really lies within these dark, crushing waters, in volume nearly 1.37 billion cubic kilometers. The hidden terrain of the deep seafloor might surprise you with its underwater mountain ranges, broad plains, hydrothermal vents and deep trenches. As we probe these depths with increasingly sophisticated tools, what we're finding is revolutionizing our understanding of this little-known, yet largest, habitat on Earth. In this book, you'll journey into the deep, looking over the shoulders of deep-sea explorers as they unravel the puzzles of life in the deep sea.

From the HMS Challenger *expedition reports.*

Architeuthis, *circa 1861.*

Amid nameless sparks, unexplained luminous explosions, abortive glimpses of strange organisms … a definite new fish or other creature.
—*Notes from William Beebe's log, 1934*

Viperfish from Beebe's bathysphere dives.

Exploring the Deep Sea

Deep Rover, *a single-seat submersible that excels at midwater research (above).*

The submersible sits like a huge, transparent sphere, lashed by cables to the ship's rolling deck. Meter-high swells rock the ship here 20 kilometers from shore; but the after deck is so stable, it's not hard to slip into the open hatch of the submersible and up into the pilot's chamber. The bucket seat feels comfortably familiar, like that of a sports car. A low panel of toggles, gauges and camera controls faces me but the surrounding view from the panoramic window rivets my attention to the action on the ship's deck.

Settling into the pre-dive routine, I begin a check of communications and instruments and call out readings over the radio. Outside, the ship's crew closes the hatch and secures it against the pressures it will resist deep in the ocean. It feels warm inside as I wait for the crane to lift the submersible from the deck and lower it into the sea. Finally, the sub rolls in the surface swell. The sunlit water here is saturated with oxygen and is full of life.

The radio crackles as I talk to the ship, and at last a voice from the bridge gives permission for the sub to descend into the twilight depths. Slowly, the submersible sinks below the surface. A few bubbles rise around us, then a calm and peaceful purpose settles in. The sea darkens from turquoise to azure, then deepest blue. I monitor our descent, reporting back to the ship, "100 meters, 200 meters, 300 meters … I can see a huge siphonophore … nearly 40-meters long."

In front of us, the colonial siphonophore *(Praya dubia)* leads with its pulsing, heart-shaped swimming bells, trailing its long ropelike body. A chain of yellowish knobs—*Praya's* constituents—line up in sequence to share the benefits of their individual tasks:

The hatchetfish, Argyropelecus, *scans the waters above it with huge eyes (right).*

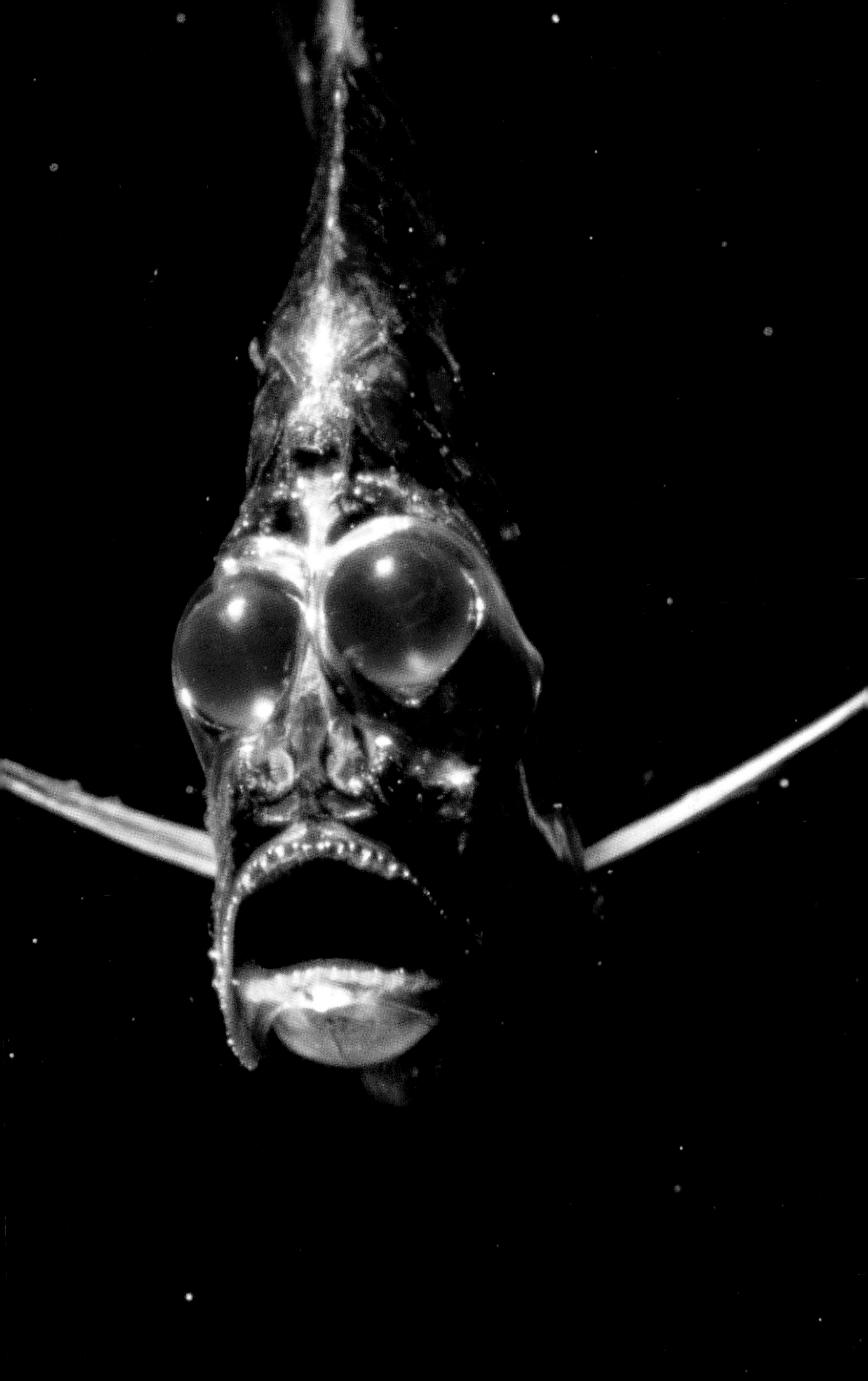

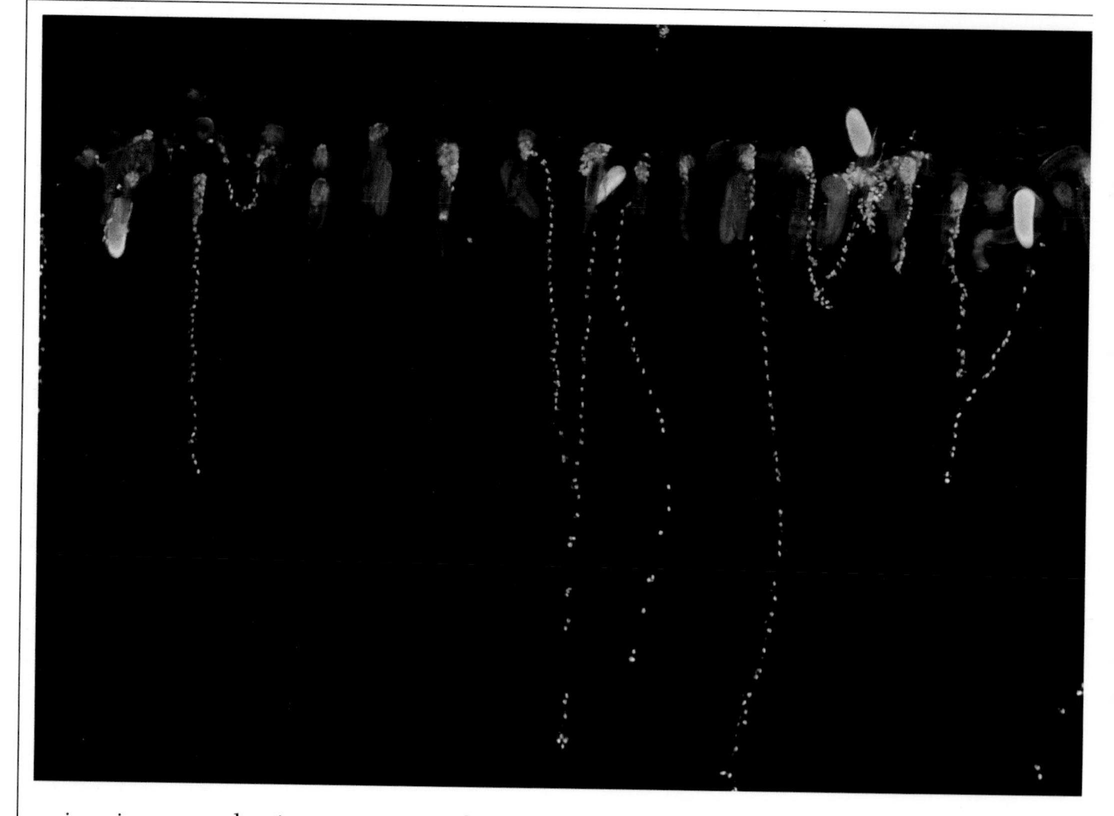

Individual elements of the siphonophore Praya.

swimming, reproduction, protection, food capture and digestion. Like a knotty drifting fishnet, this slender *Praya* chain has many swollen stomachs filled with krill and tiny shrimp.

The submersible continues to descend, revealing distinct layers, from nearly opaque water filled with microscopic particles, to regions of crystal-clear visibility. Some layers persist year-round; short-lived bands disintegrate after a day. Like habitats on land, there is a sort of topography in this midwater region, but it is fragile terrain—a dark world filled with ragged, three-dimensional spidery webs.

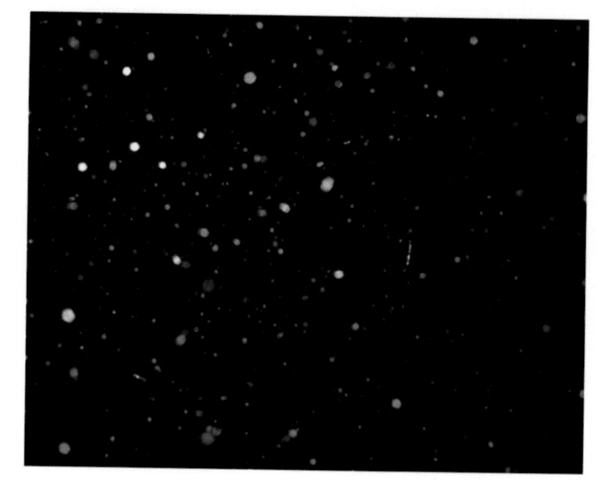

Particles of marine snow occur throughout the water column, and provide food for a wide range of pelagic and benthic animals.

The alternating layers of water contain sparse and dense concentrations of marine snow—the drifting flecks of cast-off parts from living and dead organisms. Plant cells, broken siphonophore tails, tentacles from jellies and the mucous feeding-webs of swimming snails convey ephemeral testimony to changes of fortune in deep ocean waters. Molted skeletons of krill and prawns form a transient part of the suspended material. Basketball-sized feeding filters from larvaceans collapse and sink quickly, joined by material sloughed off from sources we don't yet know. Drifting vertically like a stubby pencil dropped from above, a worm *(Poeobius meseres)* with greenish tentacles clings to some slowly sinking fragments.

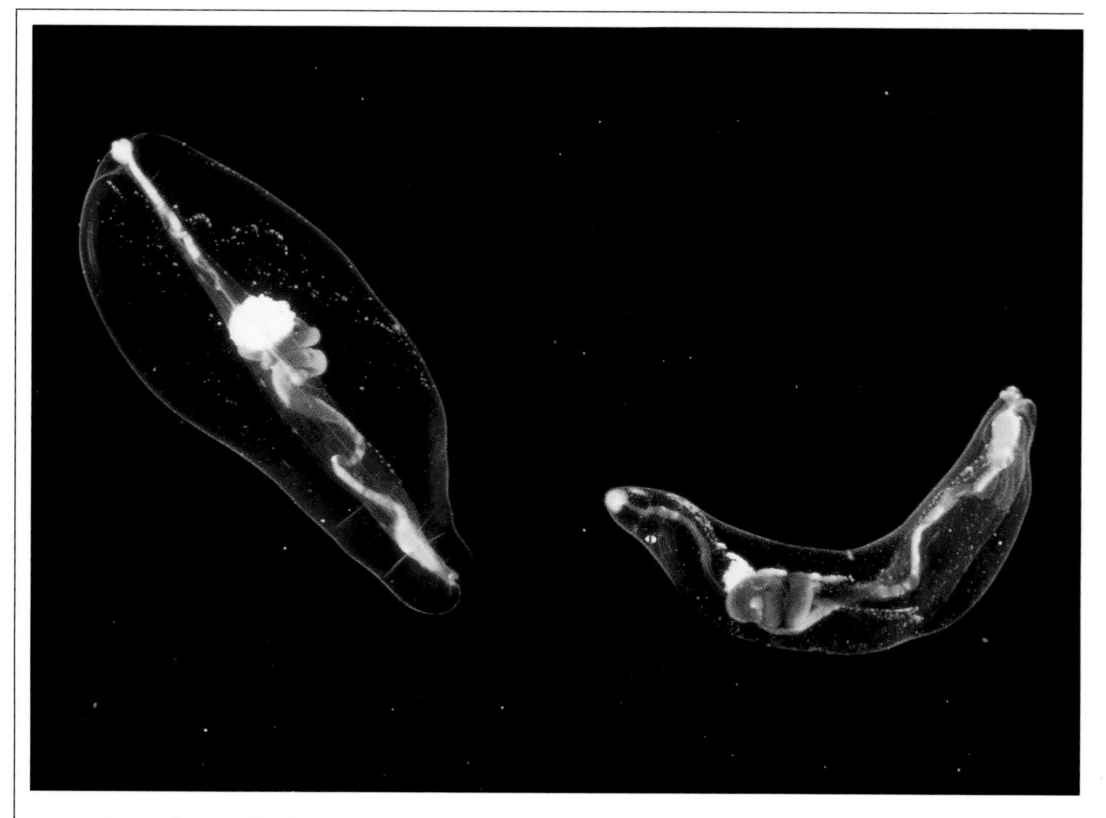

Poeobius is a lightweight midwater worm that feeds on marine snow.

A short chain of gelatinous salps captures bits of food by slowly pushing water through their bodies. Tiny crustacean-grazers scull in place, creating their own feeding eddies. Owlfish *(Bathylagus milleri)* with large, unblinking eyes drift by with heads pointed down, ever-poised to feed or flee. The general picture of calm is punctuated by random bursts of action—as a fluid school of market squid *(Loligo opalescens)* jets through, disappears, then bursts back through the scene.

The movements of the squid trigger a chain reaction of resident luminescence in the water. I shut off the submersible lights to focus on the display, which propagates like an echo through the water. Fishes, ready for action, move in for a closer look when the background lights up, stirring up even more flashes. The layers of illuminated particles resemble heat lightning rippling through a cloudy summer night.

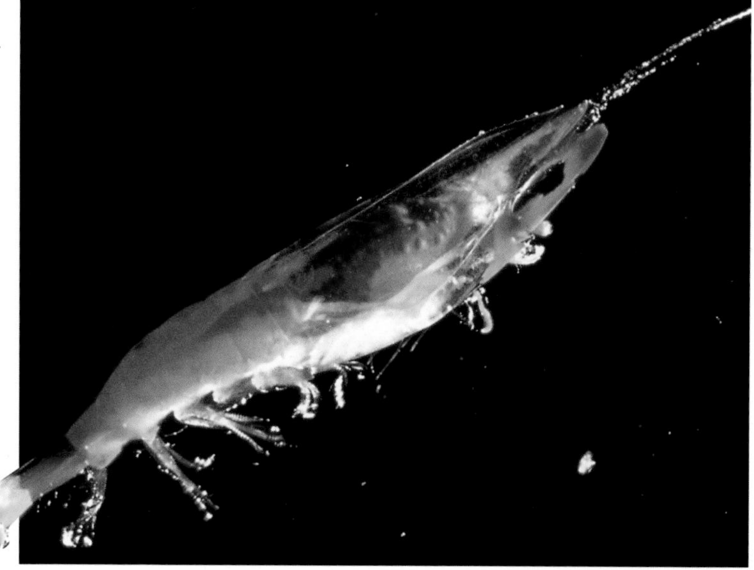

A mesopelagic shrimp.

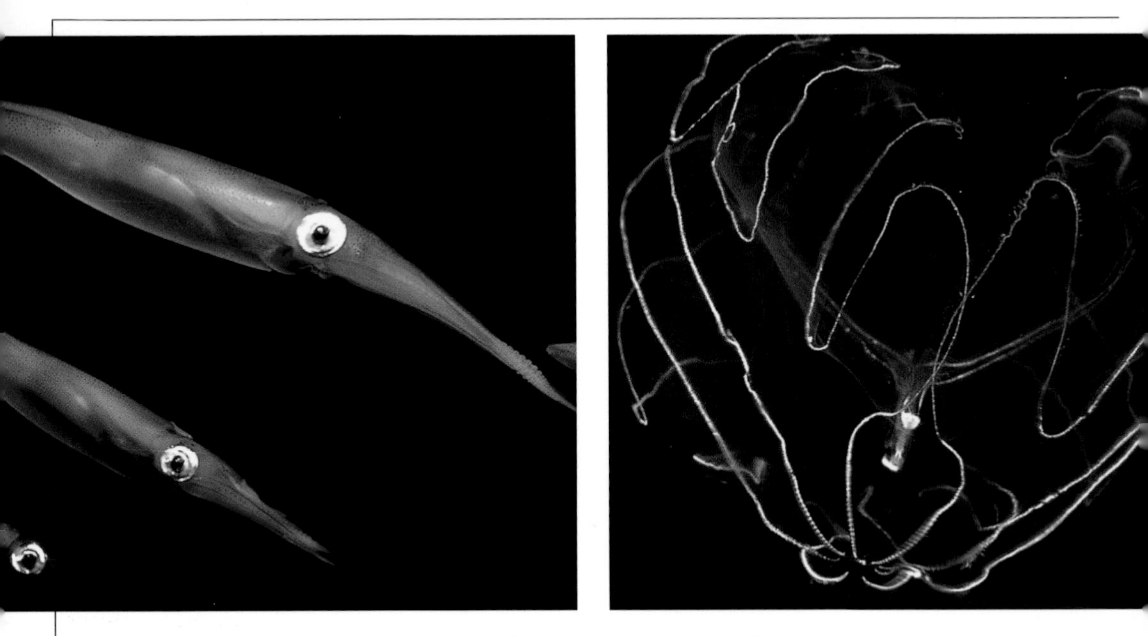

Young market squid, Loligo, *begin schooling soon after hatching (left).* Bathocyroe *is a lobate ctenophore (above).*

We descend to 700 meters, where the oxygen content of the sea slips to less than 10 percent of the level in the saturated surface waters. With the submersible lights off, the world seems inky black, cool and quiet. Only flickers of light from unseen glowing creatures and marine snow break the darkness. Few animals can stand the physiological contraints of near anoxia in this region. In water, as in air, life usually craves its source of oxygen. Some crustaceans are more tolerant of low-oxygen levels, including large mysids and some copepods. A ferocious-looking fangtooth fish *(Anoplogaster cornuta)* drifts through, virtually immobile.

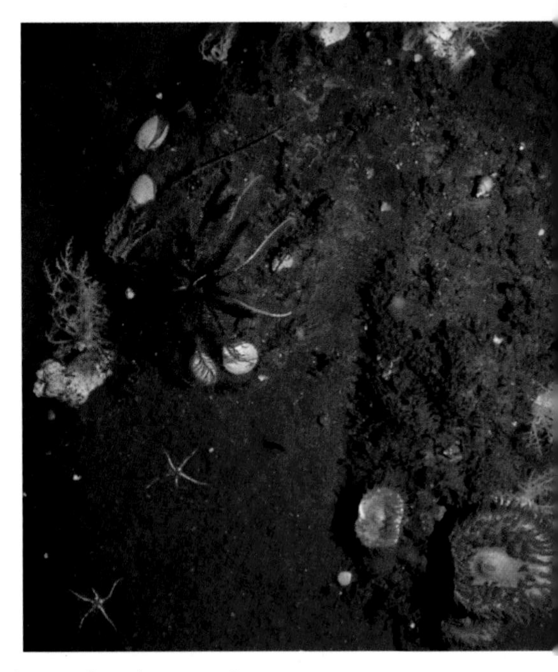

The wall of the Monterey Canyon.

Below this zone, the submersible sinks through a sharp transition from relatively clear water to a milky layer of very tiny particles. The milky layer indicates an abrupt increase in oxygen and at 1,000 meters the oxygen concentration is higher—five times what it had been at 700 meters.

We've passed 1,200 meters; we're coming to the bottom. I scan the darkness below with ineffective eyes and barely see the cloudy outline of the seafloor coming into view. Will this dive reveal a new cold seep, a new species of life or animals acting out some unforeseen behavior to attract a mate? Then I flip the switch to the submersible lights and the sudden glare, like a surprise party, reveals a vision of delights: vertical rock walls with ornate, spongy vases a meter tall, ropy whips of soft coral and strange darting fishes. I feel a thrill at the strangeness of this new environment. Then I get to work.

Monterey Bay Aquarium Research Institute

In 1987, David Packard founded MBARI, an independent, sister institution to the Monterey Bay Aquarium. MBARI's mission is to develop new technologies for deep-ocean research and to use these new tools to conduct studies of important scientific questions that could not be addressed without advanced technology. MBARI has focused on the development of remotely operated vehicles (ROVs). MBARI's first ROV, the *Ventana*, began operating in 1988 and in its first decade made more than 1,400 dives into the waters of Monterey Bay, a record that far surpasses that of any other scientific ROV.

MBARI's ROV Ventana *during launch from the R/V* Point Lobos.

Ventana is a large and powerful ROV. It stands 2.1 meters high, 1.5 meters wide, with a length of 3.3 meters and a weight of 2,338 kilograms. It has a 40-horsepower hydraulic system that drives six thrusters to give it mobility in any direction. It has two manipulator arms, a broadcast-quality video camera, lights and a suite of instruments, tools and samplers that can be configured for specific missions. The vehicle operates at the end of a flexible cable which transmits power down from the surface. The cable also has optical fibers in its core which carry signals from the camera and instruments up to the surface, as well as control information down from the mother ship. At the surface, pilots and scientists sit side by side in a control room aboard the research vessel *Point Lobos*. They face a large bank of screens and instrument readouts which allow them to monitor and control every aspect of the vehicle, as well as observe and conduct manipulative work in the deep-sea environment.

Ventana has a depth rating of 1,850 meters. For deeper investigations we have continued to work with manned submersibles. In 1989, we used *Alvin* to explore the floor of the Monterey Canyon down to 3,000 meters, successfully searching for pocket communities of animals that arise where water carrying hydrogen sulfide seeps from the seafloor. In 1990, after the large Loma Prieta earthquake, we used two Russian submersibles, *Mir I* and *Mir II*, to examine the scouring effects of turbidity flows (like undersea flash floods of sediment) on the animal communities that inhabit the submarine canyon's walls. In 1997, MBARI's engineers and operations personnel launched the ROV *Tiburon*, which will carry our deep-diving research into the next century.

Tiburon is a new generation ROV, designed specifically for scientific research. It has a 4,000-meter depth range and carries many technological innovations based on our years of experience with *Ventana*. It has twin cameras that allow the pilots and scientists to make their observations independently, as well as providing a panoramic view. It has a variable ballast system that can trim the ROV to neutral buoyancy at any depth. Its thrusters are quiet, driven by electricity instead of hydraulics, and overall the vehicle is designed to be as unobtrusive as possible. In its initial deployments we have used it to investigate the vertical migrations of midwater animals, to study the effects of El Niño on changes in the composition of the midwater community and to locate new seep communities. *Tiburon* has set a series of new diving records for MBARI, including one excursion to 4,004 meters, surpassing its design depth.

Deep Sea Players

MBARI's 4,000-meter ROV Tiburon *during launch.*

The deep sea is the largest unexplored region of the earth. Today, we are in a new age of discovery with new technological resources and expertise devoted to scientific investigation of the ocean's depths. The depths are literally uncharted territory, so one thrust of science and technology is to map the ocean floor. Around the world scientists are also conducting basic research on the ecology of the midwaters, and on the geology, chemistry and biology of the deep seafloor. Special emphasis is placed on the active parts of the seafloor including hot vents, cold seeps, ridges and volcanoes. Because journeying into the deep sea presents formidable physical challenges to scientists, the development of vehicles and instruments for research is critical.

In Monterey Bay, scientists and engineers at MBARI focus much of their research on the Monterey Canyon, making several trips a week to explore the canyon just offshore. MBARI is a key player in revealing midwater ecology, using high-quality video as a primary tool. Cameras and other instruments are mounted on the remotely operated vehicles, *Ventana* and the new *Tiburon*. Biologists, geologists and chemists collaborate with engineers in designing tools to measure, collect and make long-term observations of the life and geology in the submarine canyon, including the cold seep communities on its flanks.

Other institutions on the West Coast which explore the deep include the Scripps Institution of Oceanography in La Jolla and the University of Washington in Seattle. Scripps and UW scientists are cooperating to establish undersea observatories off Oregon and Washington. These consist of instrument packages that collect data around the clock, recording changes in deep seafloor activity and conditions. University of Washington researchers are also focusing on the biology of hot vents, investigating new species of heat-loving microbes. Scientists at Scripps used sound waves to analyze the structure of the Earth's crust beneath the mid-ocean ridge in the eastern Pacific. Other scientists at Scripps are looking at the relationship between organic matter sinking from the surface and the diversity of animals on the seafloor. Recently, Scripps scientists produced the first satellite-based map of the world's seafloor. Before this effort, we had better maps of the moon than of the Earth's ocean floor.

Several institutions on the East Coast are conducting innovative deep-sea research as well, among them, the Woods Hole Oceanographic Institution (WHOI). WHOI is the home of the most famous crewed submersible, *Alvin,* used by researchers from around the world. WHOI's Deep Submergence Laboratory's mission is to develop systems for remote exploration of the deep seafloor. Their remotely operated vehicles and instruments include the ROV *Jason* and towed instrument sleds used for detailed surveys of the bottom. Among other deep-sea investigations, researchers at WHOI are also focused on the biology and geology of the vent systems of the Mid-Atlantic Ridge.

In Florida, the Harbor Branch Oceanographic Institution (HBOI) has three crewed submersibles, the *Johnson-Sea-Link I* and *II* and *Clelia*. Scientists use these submersibles to study the ecology of midwater and benthic animals. Other HBOI scientists are investigating bioluminescence and the visual ecology of deep-sea organisms. Working with engineers, HBOI scientists have developed many advanced instruments for making in situ measurements.

There are a number of deep-sea research efforts around the world. In Japan, the Japan Marine Science and Technology Center (JAMSTEC) conducts investigations in midwater as well as on the seafloor. JAMSTEC has an array of vehicles including the *Shinkai* 6500, the world's deepest-diving crewed submersible, and *Kaiko*, the deepest-diving ROV. Scientists are conducting long-term observations of the seafloor with special emphasis on hot vent and cold seep communities. They are also pioneering efforts to understand the hot vent microbes.

In Europe the French institute, IFREMER, has an effort to explore the deep sea along with other countries of the European Union. IFREMER uses the crewed submersible *Nautile* to explore the Mid-Atlantic Ridge and has developed a new, 6,000-meter ROV called *Victor*. In the United Kingdom, scientists at the Southampton Oceanography Center are documenting a rich biodiversity on the seafloor in abyssal regions, once thought to be a desert. Other researchers use the Russian *Mir* submersibles to study the geology and biology of the hot vents on the Mid-Atlantic Ridge.

Shinkai 6500, *the world's deepest-diving crewed submersible, during recovery.*

Early Explorations

The exploration of the deep sea is a fascinating journey that requires extraordinary efforts and technology. Deep-sea biologists usually mark the years 1872 to 1876 as the birth of their field of science. These were the years that a British naval corvette, *HMS Challenger*, sailed on a three-and-a-half-year voyage around the world to explore the depths of the world's oceans, examine the physical and chemical conditions and describe whatever life could be found. Prior to the *Challenger's* long cruise, many scientists believed that conditions in the deep sea were too harsh to support life. Others believed that the deep would prove to be a refuge for ancient animals, driven into the depths by modern species that now occupy the warmer and more productive shallow waters. By the time *Challenger* returned to England, her scientists had found abundant and fascinating life in deep-sea basins, wherever they looked. While more than 4,000 new species were described from the expedition, living fossils were the exception, not the rule.

HMS Challenger *during its global expedition of discovery.*

Challenger plied the seas relying on steam power as well as sails. Her steam engines drove the ship while she pulled nets through the water and dredges along the deep seafloor. Steam also powered the winches that reeled in the miles of hemp rope used to deploy the sampling gear. The nets and dredges delivered to the scientists waiting on *Challenger's* deck a wondrous array of creatures. Most were familiar types—fishes, crabs, sea stars, worms—but the bizarre shapes of their bodies and the differences between these deep-dwellers and their shallow-water relatives were often fantastic. Among the strange animals

Deep-living benthic invertebrates.

Challenger brought to the surface was the first anglerfish (Ceratiodea) ever retrieved from its deep habitat. An angler "fishes" for its food by using a luminous lure at the end of a stalk attached to its forehead. Prey attracted to the glowing lure are sucked in as the angler suddenly opens its huge mouth. The light in the angler's lure comes from colonies of bacteria which the anglerfish provides with shelter and nutrition.

Beyond the discoveries of abundant life and the thousands of species new to science, *Challenger's* results revealed many of the basic patterns of life in the deep sea and their relations to physical properties like temperature and currents. Their surveys sketched on the globe the broad outlines of deep animal distributions and the vertical depth zones they occupy. The *Challenger* expedition established the foundation of deep-sea research for the next century and much of what we know today was first learned as a result of *Challenger's* pioneering work.

Deep-sea exploration continued after the *Challenger's* success, with major expeditions by France, Russia, Monaco, Sweden, Norway, Germany and the United States. The last of the great deep-sea expeditions ended on June 29, 1952, when the Danish ship *Galathea* returned home to Copenhagen after circumnavigating the globe for 20 months. These efforts confirmed and expanded *Challenger's* findings about the widespread distribution and broad diversity of deep-sea life. Soon after World War II, the character of deep-sea investigations changed from large-scale expeditions to shorter-term investigations focused on smaller areas.

How we understand things depends to a large degree upon the way we learn about them. Drifting over the top of a forest in a

This deep-sea anglerfish, Caulophryne, *has extremely long fin rays and very small eyes. Living in the dark, it uses the long rays to increase its sensitivity to movements in the water around it.*

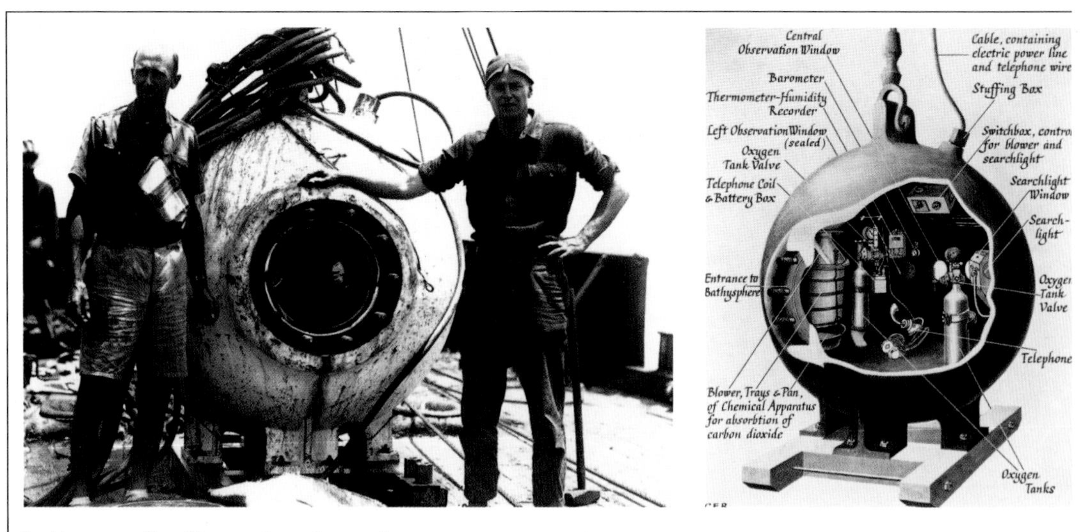

Beebe and Barton test the bathysphere in 1930 (left). Cutaway view of the bathysphere (above).

balloon will tell us a lot about the upper canopy but to investigate the forest floor, we would be much better off down on the ground, walking among the trees. The same principles apply to studying the deep ocean. For nearly a century, scientists were stuck on the decks of ships at the sea surface, using nets and dredges to grope blindly into the waters below. Trying to understand the largest animal communities on Earth, based on the dead and dying specimens hauled up and dumped into a bucket, is a bit like reaching into a bag of donuts and trying to tell the lemon from the chocolate without looking or tasting.

The first scientist ever to enter the deep sea and observe its residents firsthand was Dr. William Beebe of the New York Zoological Society. In 1930, Beebe began testing a new technology for deep-sea exploration. The device, called the bathysphere, was a hollow steel sphere, 144 centimeters in diameter, with walls about four centimeters thick. The bathysphere was lowered at the end of a steel cable from a barge at the surface. Alongside the support cable was an electric cable that carried power and a telephone line. The sphere had two windows made of fused quartz, the only transparent material, then available, strong enough to withstand the pressures they faced. Inside, oxygen was bled slowly from two tanks, carbon dioxide was absorbed by a tray of soda lime and air was circulated by hand with a palm-leaf fan.

Dr. William Beebe was the first person to see deep-sea animals in their natural habitat, and come back.

Over the course of four years, the bathysphere made 32 dives, most of them with Beebe and engineer Otis Barton inside. Off Bermuda on August 15, 1934, they reached a maximum depth of 3,028 feet, more than 805 meters. With an arc light shining out one window and Beebe's face pressed against the other, the interior of the ocean was revealed for the first time. One of the most impressive things Beebe saw was the widespread use of bioluminescence by midwater animals. The displays of light were like fireworks against the dark water. One small red shrimp squirted out a cloud of bright blue luminescence when threatened by predators. In the dark, the cloud of light shielded the shrimp in the same way that

The bathysphere returns to the sunlight (left) and Beebe emerges (right).

an octopus's ink cloud protects it in well-lit waters.

Photography from the bathysphere was not successful and the only way Beebe could visually document his observations was to describe them to an artist. The resulting series of paintings by Else Bostelmann is a vivid and beautiful record of these first glimpses of life in the depths. One of my favorites shows the bathysphere looming out of the darkness with its searchlight shining brightly out of the window. Around it are two, sinuous, dark-bodied dragonfish (Melanostomiatidae) of immense size. These giants have never been seen again.

The Evolution of Deep-Sea Vehicles Beebe's revolutionary approach to deep-sea science and his remarkable findings were not well received by the rest of the scientific community. It was not until after World War II that scientists again began to enter the deep sea. The trend began slowly, and with a few false starts. In biology, we recognize a process called coevolution, where two species evolve, influenced by each other with mutual benefit. Flowering plants and their insect pollinators are a good example. In science, we also see a coevolutionary relationship between technology and the kinds of research that it enables. The development of undersea vehicles allowed scientists to enter the deep-sea habitat, a step which revolutionized the field. In turn, scientists influenced the design of subsequent generations of vehicles to better suit their needs.

The next step in the evolution of undersea vehicles after the bathysphere was the bathyscaphe, an underwater version of the high-altitude balloons that its designer, Auguste Piccard, built for studying the upper atmosphere. Piccard replaced the balloon's gas bag with a gasoline-filled float and the crew gondola with a pressure-resistant personnel sphere. Georges Houot of the French Navy dove the bathyscaphe in the Mediterranean and the Atlantic ocean, ultimately reaching a depth of 4,000 meters. The dives focused on midwater and benthic biology. Among the findings were tripod fish (Bathypteroidae) that use their long pelvic and tail fins to stand above the seafloor to intercept food swept along by currents. Scientists also learned that fishes called barracudinas (Paralepididae) orient themselves vertically so that they can see the shadows of their prey silhouetted against the lighted surface waters.

A second-generation bathyscaphe, called *Trieste,* was launched in 1953. Purchased by the U.S. Navy in 1958, she began a research career that included detailed studies of the vertical migrations of midwater animals and the fauna of the deep seafloor off Southern California. In 1960, *Trieste* carried Don Walsh and Jacques Piccard (Auguste's son) to the floor of the Challenger Deep in the Mariana Trench—at 10,910 meters, the deepest-known spot in the ocean. While their stay on the bottom was brief, Walsh and Piccard saw a flatfish that slowly swam along the bottom and out of the light field. This was clear evidence that life thrives even in the deepest parts of the ocean.

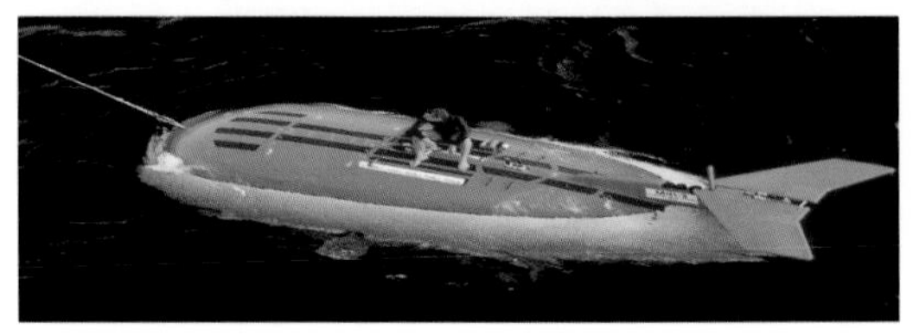

Trieste *(top),* Mir *(above), and old reliable*—Alvin *(below).*

After the bathyscaphe, a new generation of research submersibles radically altered the way science was conducted in the deep sea. The most outstanding example of these is *Alvin,* a three-person vehicle first launched in 1964 and still in service today. *Alvin* holds a pilot and two observers within a titanium personnel sphere, two meters in diameter (a significant improvement over Beebe's cramped quarters). The strong hull allows dives down to 4,500 meters. Three acrylic windows provide visibility and several pressure-tolerant lights are mounted on the outside. *Alvin* has two mechanical arms, called manipulators, and a platform mounted forward and underneath the central window for samplers, tools and instruments.

One of the biggest improvements *Alvin* has over its predecessors is mobility. Three thrusters on the tail drive the sub horizontally, and two smaller vertical thrusters on the sides take it up and down. Top speed is about two knots and the propulsion system gives *Alvin* the ability to move very

Else Bostelmann: Artist of the Deep

The deep-sea watercolors of artist Else Bostelmann are the impressive product of a unique combination–art linked by technology to science. Based on descriptions by William Beebe, who was suspended hundreds of meters below the surface of the sea, Else Bostelmann painted the drama of deep-sea life.

In 1929, illustrator Else Bostelmann joined William Beebe's deep-sea research team in Bermuda and began to paint the animals dredged up in the team's nets. She meticulously chronicled rare animals as they floated in laboratory trays and often worked in a darkroom to capture the luminescence of these fishes.

When Beebe began his deep-sea dives in the bathysphere designed by Otis Barton, Bostelmann's role became even more important. On the sea's surface—at the other end of a phone line—Bostelmann listened carefully, taking notes on the creatures from below as Beebe described them. Later she painted what Beebe had described. Together, Bostelmann's paintings and Beebe's writing captured the wonders of deep-sea life.

Before leaving the research team in 1934, Bostelmann created 50 paintings of rare and often never-before-described animals. Until her death in 1961, she continued her career as an illustrator, particularly of plants, and eventually illustrated 14 children's books. Today, most of her original deep-sea paintings are stored in the archives of the Wildlife Conservation Society in New York.

Else Bostlemann translated Beebe's verbal descriptions of midwater animals into wondrous images.

precisely along the bottom for detailed observations and for performing manipulative tasks.

Inside the sphere, *Alvin* is pretty cramped. There are big racks of instruments and controls, a life-support system, communications gear, still cameras, tape recorders, notebooks, sweaters, lunch and more legs, it seems, than three people ought to have. Scrunched up against the window, though, you forget about everything but the thrill of exploration. *Alvin* led a revolution in deep-sea research by changing the perspective of the scientists from remote to direct. No longer were they stuck at the surface on the deck of a ship high above their work. Despite the advantages conferred by *Alvin* and other research submersibles, the great majority of deep-sea biology is still conducted the old-fashioned way—with nets and dredges dragged behind a ship at the surface.

In 1985, along with a group of colleagues, I used the one-person submersible *Deep Rover* to study animals living in the upper 1,000 meters of the water that fills the Monterey Canyon. We made 55 dives over a month-long period and brought back specimens and video images that radically altered our perceptions of deep, midwater biology.

Deep Rover is a small submersible built around a transparent, acrylic sphere that holds the lone scientist/pilot. Sitting upright in a contoured seat, I had an unobstructed, panoramic view of everything around me. Fitted with four thrusters, the vehicle has great mobility in all directions. Cameras, lights, samplers, instruments and two manipulators are mounted outside the personnel sphere, all controlled by ergonomically intuitive switches inside. Unlike *Alvin* and most other research subs designed to work on the bottom, *Deep Rover* can be trimmed out to neutral buoyancy, which gives it great freedom to explore the midwaters.

The evolution of undersea vehicles beyond *Deep Rover* has led to significant changes in the way deep-sea biology is conducted. In the last decade, a new class of undersea vehicles has begun to carry out research in the deep sea. These are the ROVs (remotely operated vehicles), which are controlled by pilots and scientists aboard mother ships at the ocean's surface. The most successful scientific ROV is the *Ventana*, operated by the Monterey Bay Aquarium Research Institute (MBARI).

MBARI is one of the research institutions ringing Monterey Bay which are at the forefront of deep-sea research. The Monterey Canyon, a huge cleft in the continental shelf that brings deep water very close to shore, allows easy access to the depths and serves as a natural laboratory for deep-sea science. In this book, we will use Monterey Bay and its submarine canyon as a window into the deep sea.

This field of research is in a period of transition, from remote to in situ technologies, and from observation to quantification and experimentation. Because of these changes, deep-sea biology has become one of the most exciting and interesting fields of science.

But to unravel the mysteries surrounding life in the deep sea, we must first understand the forces at work there.

ROV control room (right) aboard the R/V Western Flyer *(far right), at work in Monterey Bay (below).*

2

Challenges of Life in the Deep

View of the Mid-Atlantic Ridge at 2,500 meters deep from the submersible Alvin.

Out beyond the edges of Earth's continents, submerged shelves break into slopes that plunge steeply downward to meet the floor of the sea. The waters that fill the vast spaces over the bottom average about 4,000 meters in depth, creating the largest living space on Earth. Least explored of any region of our planet, the deep waters of the ocean loom alien, inhospitable and mysterious, yet more animals live in the deep sea than in any other habitat on Earth.

Satellite data are used to generate maps (right) of sea-surface temperatures, to show springtime upwelling (top) and warmer conditions in autumn, off Monterey (bottom).

For the creatures that thrive in the deep, three physical factors dominate their lives: light, temperature and pressure. The deeper you go, the farther you get from the sunlight that bathes the surface of the sea. By the time you reach a depth of only 150 meters, 99 percent of the light reaching the surface has been absorbed. Light suitable for photosynthesis by planktonic plants at the base of the food web only reaches about five percent of the way down to the deep seafloor, confining phytoplankton to the very top layer of the ocean. The vast majority of the ocean is profoundly dark.

Sunlight warms the top layer of the ocean but because it is so quickly absorbed, most of the deep sea is cold, with temperatures only a few degrees above 0° centigrade. And the deeper you go, the greater the weight of the water above you. This translates to pressure. At the ocean's average depth, the pressure is nearly three-tons-per-square-inch.

3-D view of the Monterey submarine canyon (above).

Besides light, temperature and pressure, the terrain and the circulation of waters in the deep sea also greatly influence what lives there, and how it survives.

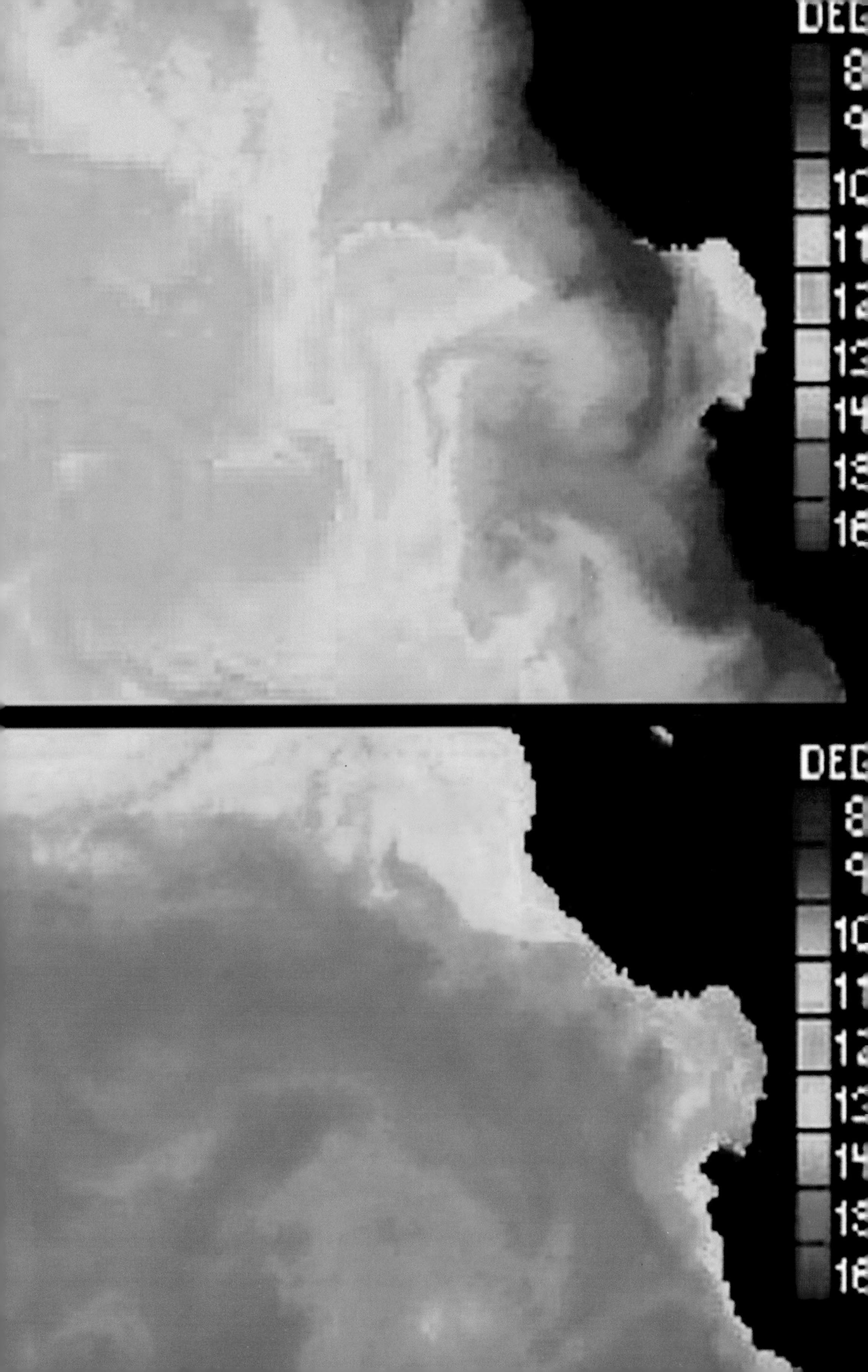
DEG
8
10
DEG
8
10

Geology of the Deep

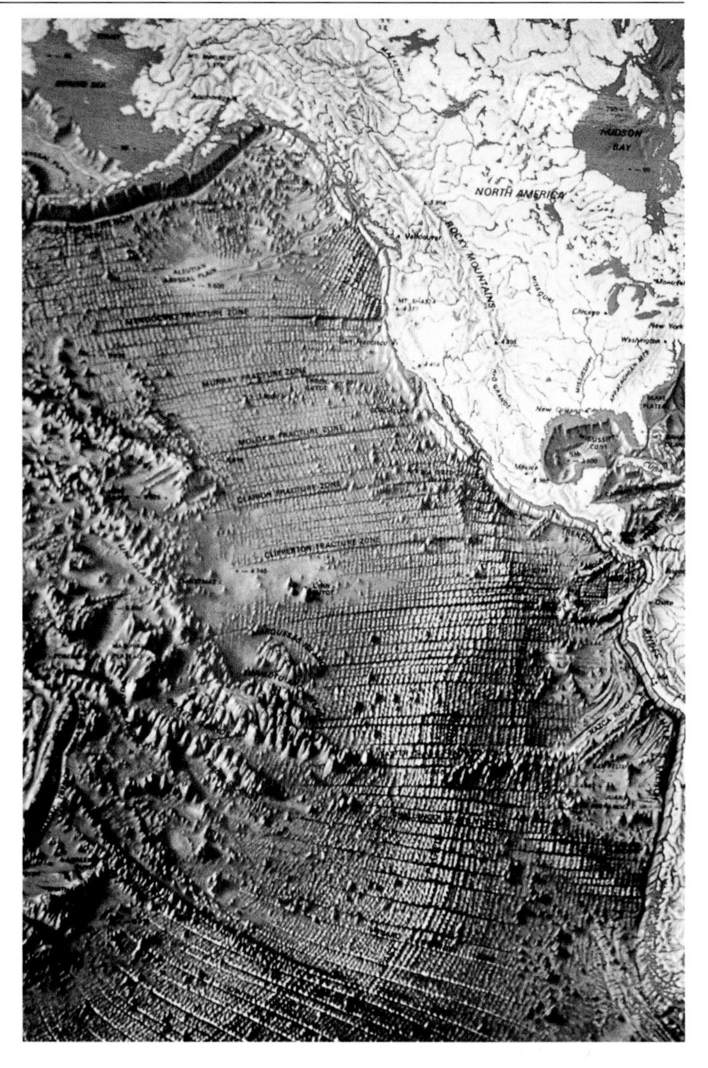

Seafloor topography of the eastern Pacific Ocean basin.

If we could drain the water out of the oceans, we would see the true geological face of the Earth. But because roughly three-quarters of the planet is covered by water, that view has been obscured. The scientists aboard *Challenger* "sounded" the oceans by lowering lead weights on long lines to measure the depths of the waters they studied. From this practice they gained a very rough idea of the shape of the seafloor below them. After World War II, "echo-sounding" with sound pulses bounced off the bottom gave a much more detailed picture. From thousands of track lines run across the oceans by ships with echo-sounders, geologists have put together a nearly complete picture of the face of the deep. More recently, satellite-borne radar has surveyed the heights of the ocean's surface, which varies slightly with the depth of the water beneath, to give us an integrated, global picture of the seafloor.

The crust of the Earth rides above the inner mantle in two basic forms, and at two different levels. The continents are made up of thicker, lighter rock that reach an average height of just less than a kilometer above sea level. The oceanic crust is thinner and much more dense, with an average depth of about four kilometers below sea level. The margins of the continents consist of flattened shelves that break seaward into steep slopes that pitch downward to the floors of the great ocean basins. The slope region is where the continental crust meets the oceanic crust.

The floors of the ocean basins are called the abyssal plains, and they are often flat or gently undulating, usually covered by a thick layer of sediments. Near the continents, sand, mud and clay particles cover the abyssal hills. These sediments originated on land but were transported to the seafloor by runoff, with the larger, heavier sand particles sinking closest to the continents and the fine clays deposited farthest out. The majority of deep-ocean sediments are composed of silica and calcium carbonate, derived not from terrestrial sources but from tiny, shell-building organisms in the plankton. The most important of these are the radiolarians, which build their shells from silica, and the foraminiferans, whose shells are calcareous.

The most spectacular feature of the deep seafloor is a chain of mountain ranges, extending from the Arctic down the Atlantic basin, into the Indian Ocean and diagonally across the South Pacific. This is the mid-ocean ridge system where new ocean crust is formed. Running the length of the ridge system's crest is an extensive, fissured rift valley, where hot, new rock is first exposed. Parallel to the ridge axis on either side is a linear series of smaller hills that represent previous surges of crust from its source at the ridge crest. Intersecting the long axis of the ridge system at right angles is a series of cracks and fractures that result from movements by the newly formed crust. The ridge system is where we find hydrothermal vents and the unique animal communities they sustain.

Japan's Kaiko, *the world's deepest-diving ROV.*

The ridge system covers about a third of the ocean floor. The relatively smooth abyssal plains are also punctuated by abundant seamounts, inactive volcanoes that rise steeply from the seafloor. Where volcanoes break through the sea surface we find clusters of islands like the Galapagos, or island chains like Hawaii.

As new crust rises from the ridge system and cools, it pushes the older crust laterally and away from the ridges. Because the Earth's crust is broken into several huge plates, this motion causes the plates to move against each other. These interactions between plates (tectonics) are among the most powerful forces on Earth; they build mountains, cause earthquakes and in one way or another affect everything living on the planet.

In the western part of the North Pacific an oceanic plate grinds against a continental plate. But instead of pushing up to build mountains, the oceanic plate is pushed downward and the results are deep fissures in the crust called trenches. The deepest of these, the Mariana Trench, is almost 11-kilometers deep. This is the site seen first by Walsh and Piccard from the bathyscaphe *Trieste* back in 1960. More recently, Japanese scientists have sent their ROV *Kaiko* back into the trench to resume the long-interrupted exploration.

Deep rock ledges covered with fine sediments.

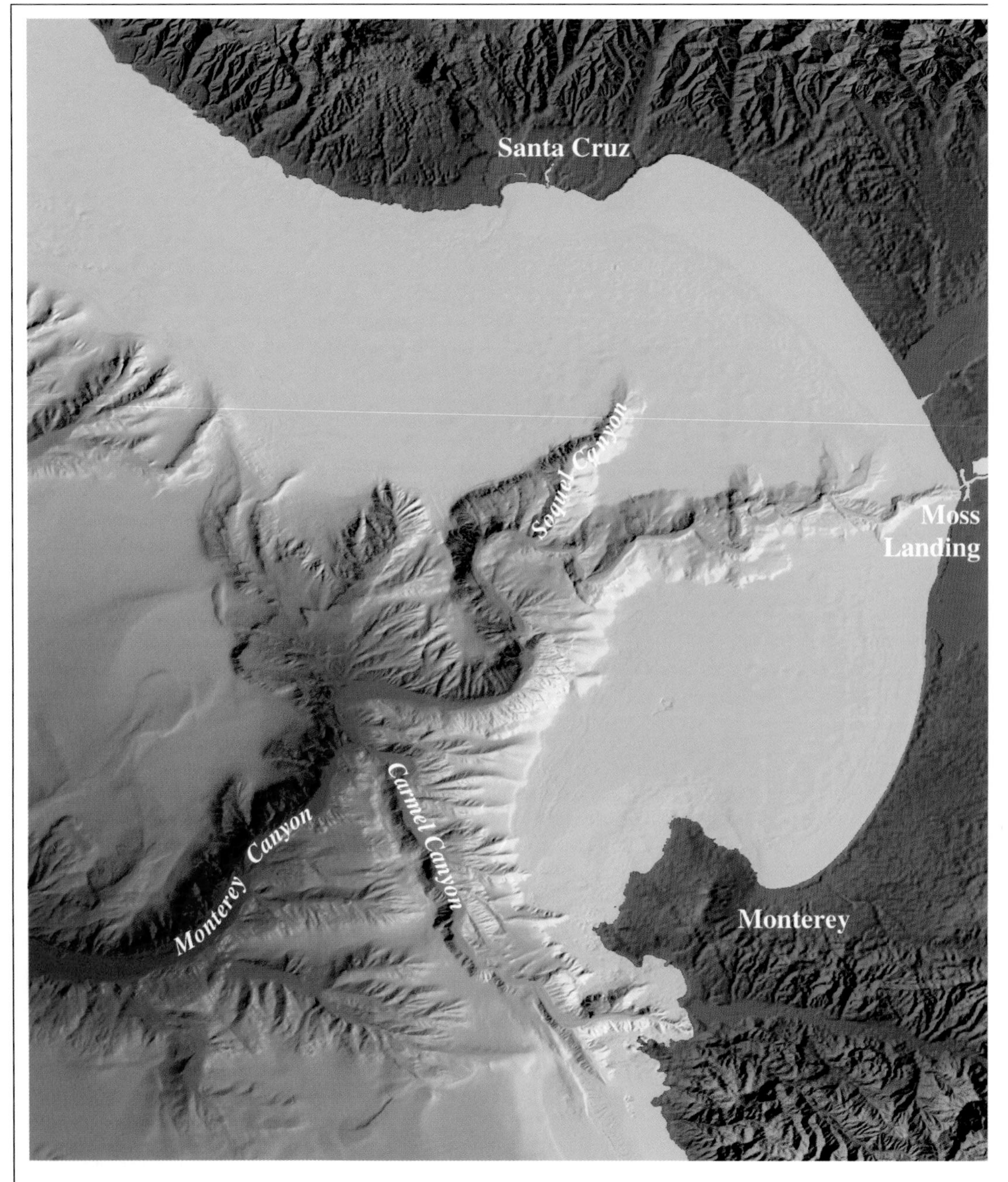

Monterey Bay and the Monterey submarine canyon system.

Submarine Canyons In some regions, continental slopes are cut through by deep, V-shaped channels called submarine canyons. They are probably caused by erosion during glacial periods when sea level is lowered. In some cases, the canyons extend up through the continental shelf, most often near river mouths. The largest submarine canyon on the west coast of the United States lies in Monterey Bay. This bay is a roughly semi-circular bite out of the central California coastline, about 40 kilometers across at its mouth and 20 kilometers from the mouth to its head. The mouth of the bay opens directly into the open waters of the eastern North Pacific.

The Monterey submarine canyon begins at the center of the bay's eastern margin and cuts through the narrow continental shelf until it reaches the floor of the North Pacific basin offshore. The scale and scope of the canyon are huge, roughly that of the Grand Canyon of the Colorado River. At the head of the canyon, at Moss Landing Harbor, the depth is less than 10 meters. At the mouth, where it breaks out into the abyssal plain, the depth is about 3,800 meters. The flanks of the Monterey Canyon slope gently from the intertidal to an average depth of about 400 meters, where they break more sharply downward as the walls of the canyon.

Along its length, the Monterey Canyon axis has several meanders, and side canyons cut through the north (Soquel Canyon) and south (Carmel Canyon) walls. As it deepens, the Monterey Canyon cuts first through layers of sedimentary rock, then granite. Near the head of the canyon are the Salinas and Pajaro rivers which transport organic-rich, terrestrial sediments into its upper reaches. Canyons are usually geologically active areas, with landslides contributing to turbidity currents that undercut the walls and cause slumping. Sediments transported down-canyon are deposited at its mouth in a broad fan.

The Monterey Canyon lies within a tectonically active region near the junction of the Pacific and North American plates. The area has many faults, both active and inactive, which contribute to the bay's topography. While the origin of this canyon is still debated by geologists, it is clear that plate motion and fault activity have played significant roles in its development. Overall, this is a dynamic geological setting and the canyon itself undergoes dramatic local changes in a relatively short time frame.

Cooled rings of lava on the wall of a collapsed lava pond on the Mid-Atlantic Ridge 2,500 meters deep.

The shallower parts of the bay and the flanks of the Monterey Canyon are generally covered with soft, fine sediments, with relatively few rocky outcrops. The floor of the canyon is also sediment-covered although rock piles and boulders are the common results of landslides and erosion of the walls. The net transport of sediments on the floor is toward greater depths to the west, and its turnover rate is considerably higher than that of sediments on the flanks. The canyon walls are steep in some areas, more gently sloping in others. In most areas, the walls consist of ridges and sediment-covered shelves which provide considerable small-scale relief.

In some parts of the canyon, tectonic activity and the presence of aquifers lead to water being squeezed from the rocks. The expressed water passes up through the sediments in some areas, and directly from rock fissures in others. In places where this water carries significant levels of hydrogen sulfide or hydrocarbons, cold-seep animal communities may become established. The seeps may be short- or long-lived but based on the estimated ages of resident bivalve molluscs, some may last for a century or more. No hydrothermal (hot water venting) activity is known to occur in Monterey Bay.

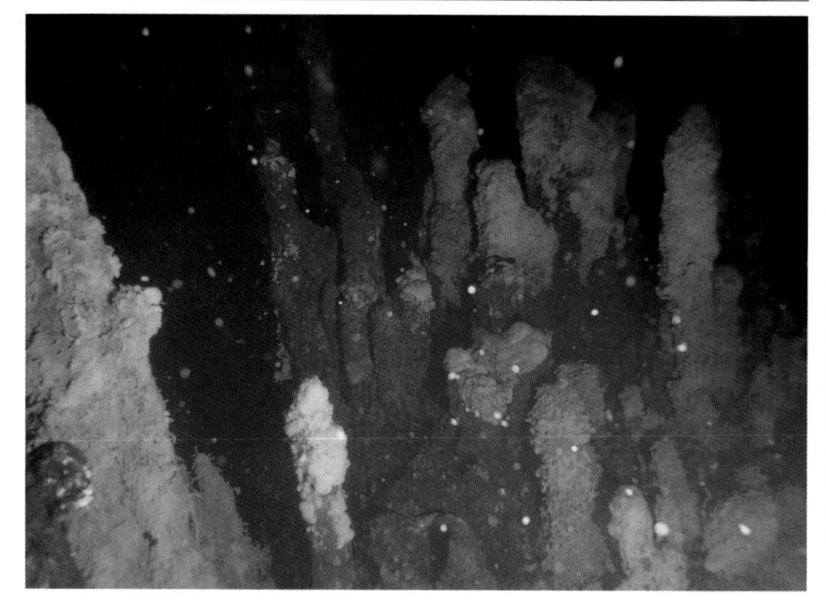

These solitary zinc-sulfide spires 2,300 meters deep on the Juan de Fuca Ridge were once active black smokers.

Cold-seep clams cluster where water rich in sulfides percolates up through the sediments (right).

Oceanography

The characteristics and variability of the waters that fill Monterey Bay and its canyon have a strong influence on the animals that live there. At the surface, the water is warmed by the sun and mixed by winds down to about 40 meters. With increasing depth its significant properties—temperature, salinity, oxygen content, turbidity—show considerable change. Temperature is relatively uniform in the mixed layer, ranging from 12° to 16° centigrade depending on the season, but it declines steadily with depth. At 400 meters the temperature is usually about 4° centigrade and it changes very little as you go deeper. Surface salinity (the saltiness of the water) is variable at the surface, depending on the amount of rain or sunshine. Below the mixed layer it remains relatively stable.

One of the most interesting features of the oceanography of Monterey Bay, and much of the eastern North Pacific, is an oxygen-minimum layer. At the sea surface, oxygen concentrations are at saturation, as you go deeper the oxygen content diminishes gradually, then rebounds. In the core of the minimum layer at 700 meters, oxygen content is less than 10 percent of that at the surface. Below the layer, at about 850 meters the oxygen level begins to rise and at 1,000 meters it is back up to about one-sixth of saturation, which holds pretty consistently to the bottom. Turbidity, or the amount of suspended particles in the water, is highest near the surface. Just below the oxygen-minimum layer is another peak of small-particle density that turns the water almost milky. In between, turbidity is highly variable with thick layers present one day, and gone the next. Just above the seafloor is another area of relatively high-particle density, due to sediments stirred up by bottom currents.

The seasonal cycle of water movements at the surface of Monterey Bay has a strong influence on its patterns of phytoplankton productivity, which is closely coupled to the food supply of the animals that live in deep water. During spring and summer, winds acting on the southerly California Current push the top layer of the water westward. This water is replaced from below in a process

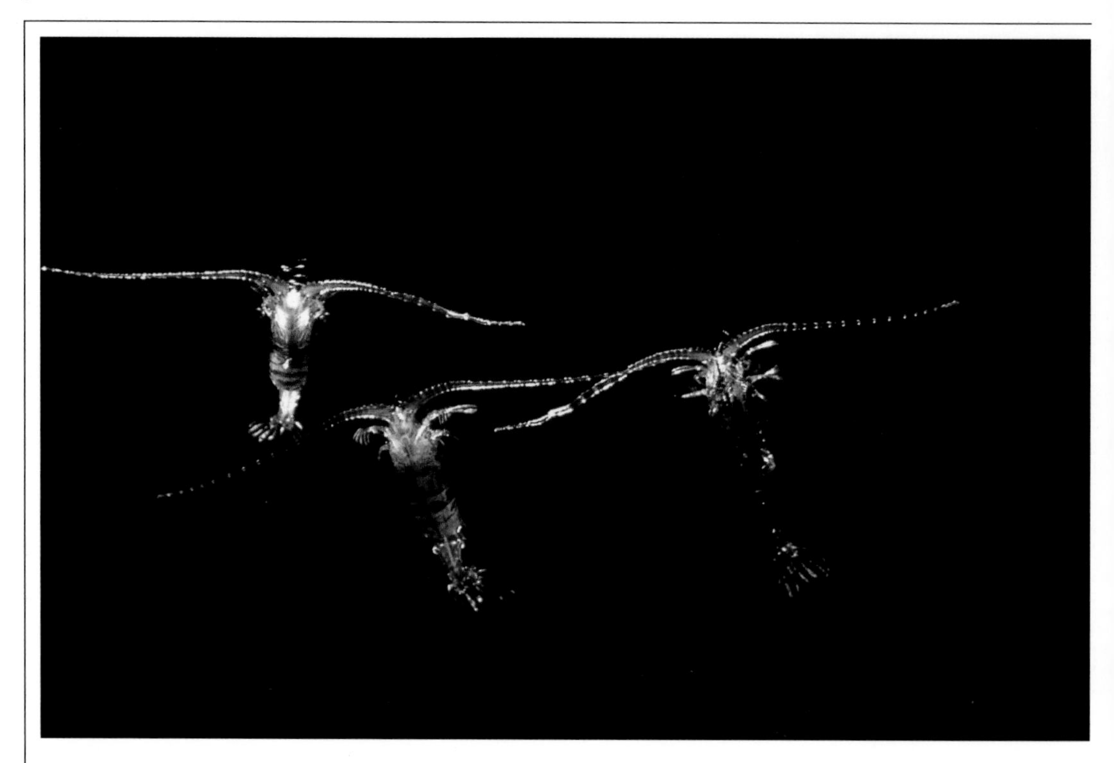

Copepods occur throughout the world ocean and are among the most important grazers on phytoplankton.

called upwelling. Upwelled water is rich in nutrients that act as fertilizer for the phytoplankton. The phytoplankton blooms last as long as this cold, nutrient-rich water is brought to the surface. Upwelling usually continues in pulses through the summer months, until the driving winds relax and upwelling shuts down.

The upwelling season is often followed by a pulse of warm, salty water from offshore that puts a warm cap on the surface of the bay. During the winter months, warm water from the south, called the Davidson Current, flows northward along the coast and dominates the surface waters of the bay. The overall pattern of seasonal oceanographic changes in the surface waters is also felt in the depths of the canyon, although dampened by the depth of the water column. Time lags of a month or more occur between the onset of upwelling and the population growth of phytoplankton-grazers like krill. Another month may pass before the predators of these grazers also show an increase.

Deep currents in the canyon generally follow its axis but cross-canyon flow has been measured as well. The deep currents flow both up-canyon and down-canyon and some of them are linked to tidal cycles. Often the flow can be vigorous. The most dramatic flows are due to turbidity currents, created when an earthquake or slump loads the water with heavy, suspended particles that are pulled down-canyon by gravity, scouring its walls and floor. While few measurements of currents in the canyon have been made, the best evidence that they are common is the presence of passive suspension-feeders like sea pens. These animals rely on currents to carry food particles to them, and the fact that they are abundant in many areas is proof that the currents exist.

Live From Monterey Canyon

Wonder what scientists see as they explore the deep? In 1989, MBARI launched a unique program that allows visitors to watch in "real time" as scientists investigate the depths of Monterey's submarine canyon. "Mysteries of the Deep: Exploring Monterey Canyon" is a multimedia presentation that uses live video from the deep and an extensive archive of deep-sea images to allow aquarium visitors to share the sense of discovery and wonder with working scientists.

As MBARI scientists investigate the canyon, images from underwater cameras on ROVs are transmitted by optical fibers up to the ship, and via microwave transmission to the aquarium's specially designed theater. During the presentation, a narrator interprets the habitats and animals on screen, answers questions from the audience and even talks to the scientists while they work. No two programs are exactly the same–one program might allow the audience to watch a marine geologist as she collects core samples from the submarine canyon wall; another might focus on a fast-moving medusa like *Colobonema*.

A special Monterey Bay Aquarium program gives the public a view of MBARI deep-sea research.

In the wake of a storm, this kelp holdfast tumbled into deep water, where it provides food for sea stars and shelter for a rockfish (above).

Lobate ctenophores, like this one, are effective predators despite their fragile structure (right).

Weather Aside from the influence of winds on upwelling, weather plays other roles in the deep cycles of productivity in Monterey Bay. Winter storms typically tear kelp plants free from their rocky anchors along the margins of the bay, and many of them wind up in the bottom of the canyon as a wind-fall food source for deep benthic animals. Likewise, heavy rain leads to organic runoff from agricultural areas and to increased turbidity from sediments carried by rivers.

Every few years, Monterey Bay experiences an El Niño event, driven by warm water that shifts along the equator from the western Pacific to the east. Uncharacteristically warm water enters the bay and dampens or delays the normal upwelling pattern. This usually diminishes the productivity of phytoplankton and the effects of diminished food are felt throughout the water column and into the depths of the canyon. The warm water also brings invasions of warm-water species from the south, which can displace the normal residents, who remain in their preferred temperature range by moving deeper or further to the north.

The conditions we find in Monterey Bay are more typical of deep coastal regions than of the open ocean, which covers most of the world's deep-sea habitats. However, because of the higher coastal zone productivity, more deep-sea animals occur on the deep continental shelves, slopes and abyssal margins than in the deep-ocean basins. Because of its natural variability, its high productivity and the deep water it brings close to shore, Monterey Bay is an ideal place to examine deep-sea biology.

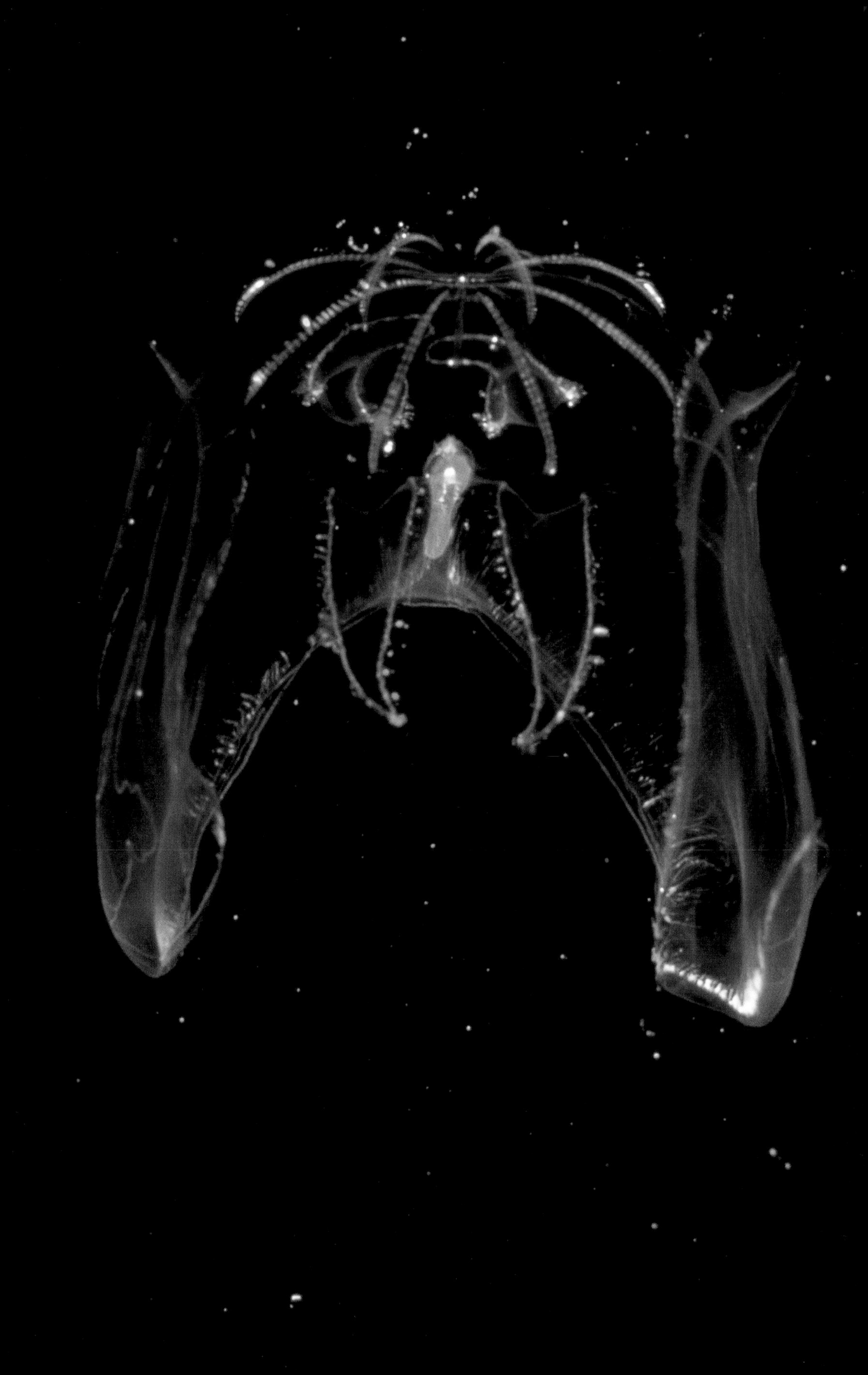

3

Life in the Ocean's Midwaters

Despite what seem to be harsh, exacting conditions, the deep sea teems with life. The modifications that have allowed life to penetrate and succeed in this strange world are fascinating examples of evolution and adaptation. We still know little about these animals because, historically, our access to the habitat has been very restricted. In recent years, new technologies have given deep-sea researchers the means to explore the depths of the ocean as never before, offering biologists windows into the only place on Earth where so many new discoveries can be made.

The foundation of the deep-sea food web lies in the sunlit surface layers of the ocean. Here, the tiny plants of the phytoplankton convert solar energy into organic material through the process of photosynthesis. Phytoplankton occur virtually everywhere in the ocean's surface waters. In some places their abundance turns the water soupy green; elsewhere they are so thinly distributed that the water appears crystal clear. The key to these differences in productivity from one area to another, or from season to season, is nutrients. Nitrogen and phosphorus compounds are the principal ingredients of this "fertilizer," just like the nutrients we add to houseplants on our windowsills. Other materials, like iron and silica, are also important to phytoplankton productivity.

The ctenophore Beroë *(above). A copepod (middle). An ostracod crustacean (bottom).*

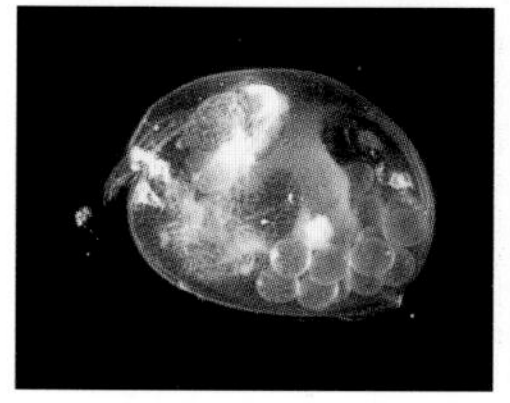

Phytoplankton growth cycles are closely linked to seasonal patterns of water movement. In Monterey Bay, for example, each spring upwelled water, rich in nutrients, reaches the surface of the bay. This water, drawn up from the depths, is cold and causes Monterey Bay's seasonal fogs. The nutrients it contains also trigger the regular spring bloom of phytoplankton that supports animal communities in deep water.

A tomopterid worm (right).

Grazers A wide range of animals graze on phytoplankton in the upper layers. Among the most abundant are two types of small crustaceans: copepods and euphausiids. Copepods are tiny teardrop-shaped animals (about the size of a rice grain), with large antennae on either side of their head and a single, centrally located eye. Copepods feed by combing phytoplankton cells and other particles out of the water with bristly appendages located near their mouths. They swim by using their five pairs of legs, but their most distinctive movements are short, quick hops generated by snapping their forked tails. These movements help them escape the attacks of predators.

Euphausiid shrimp, or krill, are an important link in the vertical food chain between the productive upper layers of the ocean, the animals that inhabit the deep seafloor, and everything in between.

Euphausiids are shrimplike grazers also known as krill, a Norwegian word for whale food. They feed on phytoplankton and other small organic particles by sieving water through a basket of bristles around their mouths. Some euphausiid species form dense, swarming groups, presumably for protection against predators. Like flying insects attracted to sources of light, krill are often drawn to the lights of undersea vehicles. Some flying insects that use the full moon as a guide for gathering in groups are fooled by artificial light sources, and it may be that euphausiids also use the moon as a cue for group behavior.

Like most crustaceans, euphausiids cast off their hard outer bodies when they grow. One clear night under a full moon, as I worked my way back up to the surface in *Deep Rover*, I passed into a thick layer of euphausiid molts. Above this was a clear layer, completely free of particles, then a dense layer of euphausiids that had recently molted. Mass molting may be like periodic rainstorms that sweep the water column of particles. Preliminary evidence suggests that molting may take place on a regular cycle, perhaps keyed by the moon. If so, then this food may be supplied to the deep seafloor from the ocean's upper layers in pulses instead of the steady rain of particles we used to imagine.

Because their food is concentrated up near the surface, copepods and euphausiids naturally do most of their feeding in the upper layers. They feed chiefly at night when the sea surface is dark. With the approach of dawn, they begin to swim downward until they reach depths of 200 to 400 meters. They spend the daylight hours in darkness, and return to the near-surface waters only as the sun sinks below the horizon. The value of this strategy is pretty clear—by using vertical migrations to remain in darkness, they reduce their chances of being eaten by predators who hunt by sight, which abound in near-surface waters.

Filter Feeders Salps and larvaceans are jellylike grazers that filter phytoplankton out of the water. Barrel-shaped salps pump water through their tubular bodies with muscular contractions that propel them as well as set up feeding currents. Salps can occur as solitary indicicuals or in chains. *Cyclosalpa* forms spiral chains that can be as large as a bicycle wheel.

While salps have internal filters and glide through the water to gather food, larvaceans build external filters and push particle-laden water through them by beating their tails. Larvacean "houses" are made of mucus and have two parts. Outside is a coarse-mesh filter to prevent large particles from clogging the fine-mesh inner filter, which selects the right size particles to eat.

Larvaceans living near the surface are usually small, and create spherical houses. At depths between 200 to 400 meters lives the giant larvacean (*Bathochordaeus charon*). Recently, we learned that *Bathochordaeus* secretes a sheet-like outer filter that grows continuously to a meter in greatest dimension. New filters are small and transparent; with age they grow and accumulate masses of particles and sediment on their upper surfaces. From a distance, some of the larger and older sheets look like floating islands with dust-covered mountains.

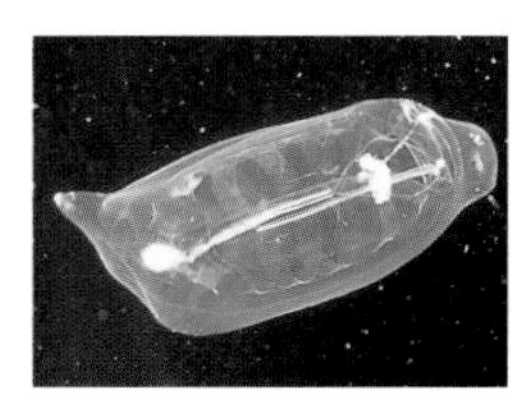

Salps (above) and larvaceans (below) feed by filtering organic particles from the water. Salps have filters within their cylindrical bodies while larvaceans construct external filters that surround and protect them.

Three other kinds of larvaceans live at increasingly greater depths in the water column of Monterey Bay: redheads (*Mesochordaeus erythrocephalus*), hammerheads (*Oikopleura villafrancae*) and an undescribed species we call "oiks." Each has a filtering structure specially adapted to best trap the particle sizes and types found at the specific depths they inhabit.

Larvacean feeding structures play two additional roles in deep-sea ecology. First, while the animal is present, its tail beat keeps the structure inflated. With time, the mucous outer cover and its accumulating detritus become oasis communities of bacteria, copepods and other small animals. Second, when the filters become clogged with heavy detritus or when a school of squid dashes through them, larvaceans abandon their houses and swim away to build new ones. The deserted structures quickly collapse without internal water flow and they begin to sink. These large, organically rich particles sink rapidly and provide an important mechanism for carrying food produced in the upper layers to the floor of the deep sea.

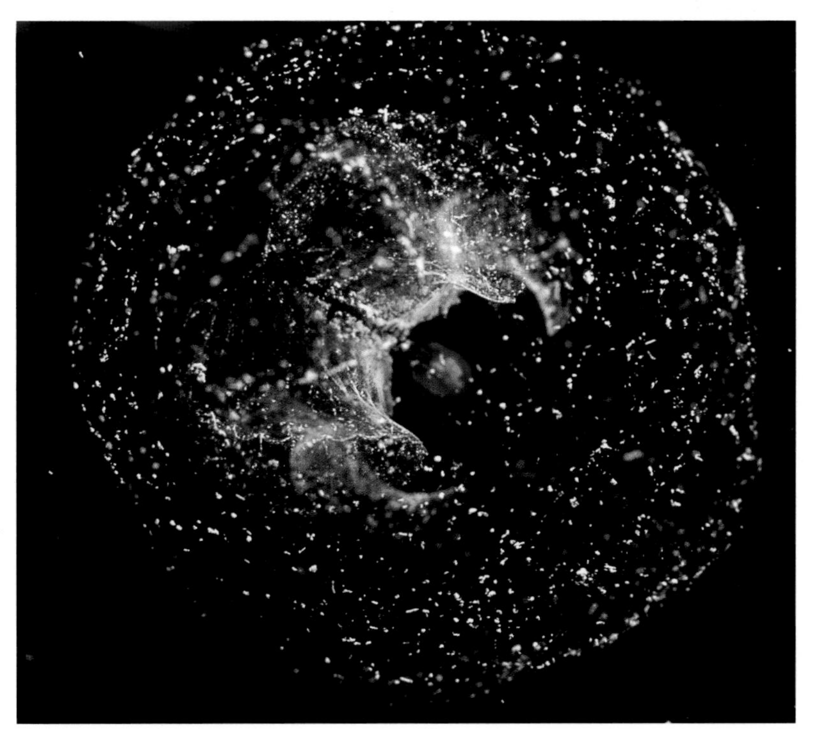

Predators Animals at the next level of the food web, beyond the phytoplankton grazers and particle feeders, are carnivores. Some of the most common animals at this ecological level are small fishes that feed on copepods and euphausiids. Among them is the silver hatchetfish *(Argyropelecus lychnus)* living at depths between 200 and 400 meters, both day and night. It feeds primarily during the day when its prey migrate downward into dark waters. To find other animals in the dark, *Argyropelecus* has a remarkable pair of eyes and an effective feeding strategy. The large, telescopic eyes are directed upward. The retina lies horizontally at the base of a silvery vertical tube with a spherical lens on top. This structure captures all of the available light coming down from above. The paired eyes are close together at the top of the fish's head, with visual fields that overlap. This gives *Argyropelecus* binocular vision, like we have, which provides excellent depth-of-field and the perception of distance for spotting prey against the backlit waters above.

To avoid falling victim to the same kind of predation from below, hatchetfish and many other midwater animals have evolved an effective counterstrategy. Along their bellies is a series of light-producing organs, (called photophores) which contain reflectors, a lens to direct the light and specialized tissue that generates light. The tissue produces two chemical compounds, luciferin and luciferase, which release energy in the form of light when they are combined in the presence of oxygen. The effect of this downward glow acts to erase the shadow cast by the fish when silhouetted against the lighted waters above.

Hatchetfish (below), lanternfish (right top and bottom) and bristlemouth fishes (right middle) have their light-producing photophores placed in characteristic positions along their bodies.

Lanternfish (Myctophidae) also use bioluminescence to conceal themselves from predators trying to spot their shadows from below. They have additional photophores distributed in specific patterns along their sides. Each species has a unique pattern, which scientists can use to tell them apart. If we can do it, then surely the fishes also use the lights to recognize members of their own species in the dark waters they inhabit. Within some species, males and females have different light organs to help identify each other.

Unlike hatchetfish, which

live at relatively constant depths, most lanternfish move up and down in the ocean. In Monterey Bay, they spend daylight hours between depths of 250 and 600 meters. When the sun sets, they migrate to the near-surface layers where they do the bulk of their feeding. These morning and evening migrations by fishes, krill, copepods and many other midwater animals, in all the world's oceans, constitute the largest mass migrations on Earth.

Further up the pelagic food web, at the fourth level, are a group of fishes and squids that feed on the fishes and crustaceans at the third level. As a rule they are not vertical migrators; instead they range widely through the ocean's upper kilometer. Some use speed to catch their prey, others use stealth and trickery.

Barracudinas are sleek, silvery fish with sharply pointed snouts and large eyes. We usually see them hanging vertically in the water, searching the dimly lit waters above for the silhouettes of fishes or shrimps. They hold position by sculling with their tail fins, then explode upward to seize their prey.

Dragonfish, such as the Pacific blackdragon (*Idiacanthus antrostomus*) have long, black bodies and huge mouths filled with long, sharp teeth. The teeth are hinged to fold backward toward the gullet but not forward, so anything that enters the mouth doesn't come out. Beneath their chins are filaments called barbels with light-producing organs and sensory fibers at their ends. Dragonfish lie

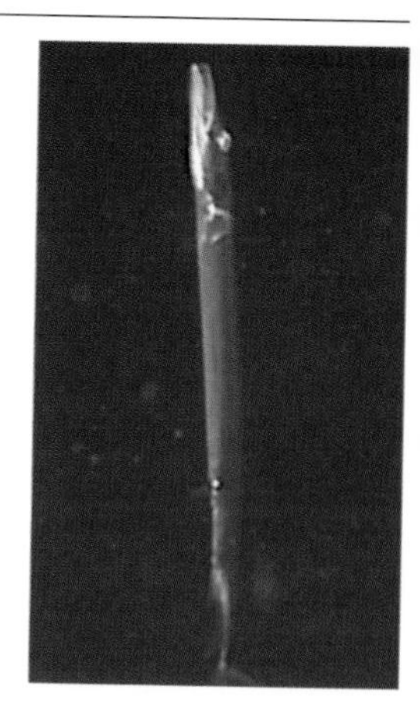

Barracudinas (above) position their bodies vertically to scan the waters above for shadows of their prey. This dragonfish (below) can eliminate its shadow with light from the photophores along its belly.

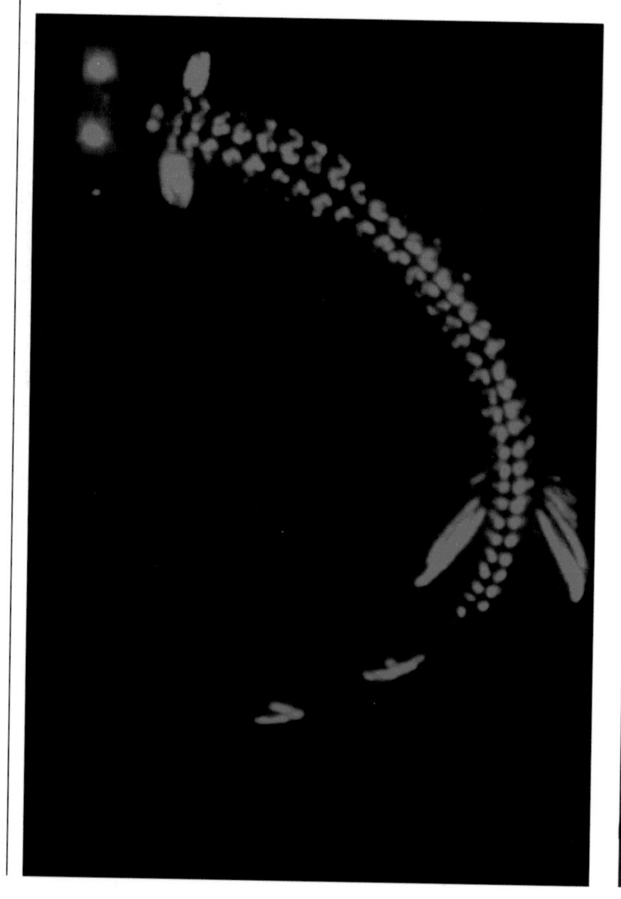

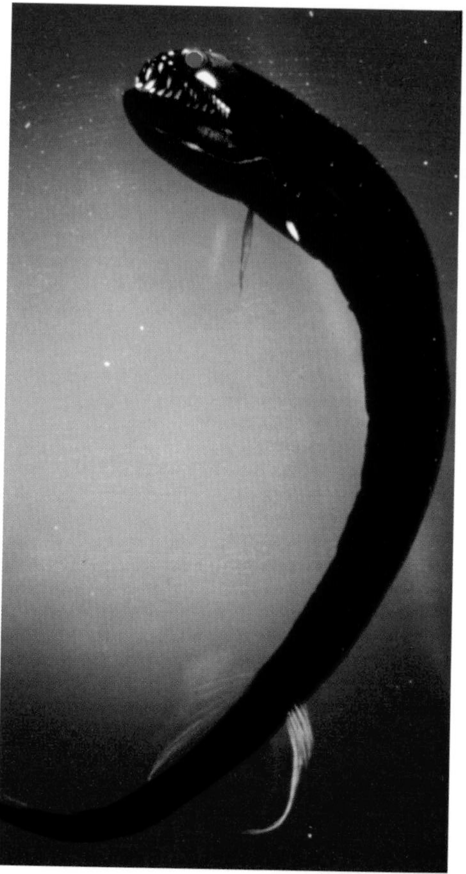

motionless, using the light organ in their barbels to attract prey. When the prey approach close enough to be detected by the sensory fibers, the predator lunges forward to grab its meal.

Chiroteuthis calyx is a squid that lures its prey like the dragonfish does. Two of its eight arms are more robust than the others, with a groove running down the center on the inside. The squid's two long feeding tentacles slide out through these grooves and over their ends to hang down well below them, when fully extended. Light organs run along the sides of the feeding tentacles. In its normal feeding posture, *Chiroteuthis* dangles the feeding tentacles down below its horizontal body, moving them up and down while flashing the light organs in a sequential wave toward the mouth. This behavior attracts fish or other prey close enough to be captured by the tentacles and drawn in to the squid's sharp beak.

Like most squid, *Chiroteuthis* and its relative, the cockatoo squid *(Galiteuthis phyllura)*, have large, well-developed eyes. While the rest of their bodies are transparent to protect them from visual predators, the eyes are so densely structured that they cannot be transparent. In order to prevent these opaque tissues from giving away their location to shadow-stalking predators, these squids have U-shaped light organs on the undersides of their eyes. Their glow eliminates the eye's shadow to an upward-looking predator. For this to be effective, the light must always be directed downward. As a result, the body of the squid pivots around its eyes whenever the animal moves.

A Pacific blackdragon attracts the fish and shrimp it eats by flashing the light at the end of its long chin barbel.

Snipe eels *(Nemichthys scolopaceus)* have long, sinuous bodies and bird-like beaks with tips that curve gracefully away from each other. The beaks are covered with tiny, hooked teeth that act like half of a velcro strip. Snipe eels sweep their beaks through the water trying to entangle the antennae of sergestid shrimps *(Sergestes similis)* which have surfaces that are like the other half of velcro. As soon as the two come in contact, the beaks lock on the antennae and the shrimp are trapped. This degree of specialization is an evolutionary gamble for the eels, because their unusual beaks are ill-equipped to capture anything else to eat. But the gamble has

The long antennae of this sergestid shrimp (left) provide physical and chemical information on its habitat. They also give this snipe eel (below) an ideal way to catch the shrimp.

paid off and as long as the abundant sergestids are present, the eels will succeed.

Hake *(Merluccius productus)* are large, aggressive, schooling fishes that cruise through the midwaters like a horde of outlaw bikers feeding on whatever fishes and shrimp they encounter. They are attracted to the lights of undersea vehicles and at times we can see their schools on our scanning sonar, extending out in front of us for thirty meters or more. They may be drawn to disturbances in light as an indication that active prey are present. While diving

Deep Rover in Monterey Bay, I was occasionally surrounded by hundreds of hake, so thick that I had to turn out the lights and dive away in order to resume my observations of the usually quiet midwater scene. Hake are a commercially valuable species in the eastern North Pacific, and their schools are hunted by fishermen in coastal waters from Alaska to Mexico.

The Jelly Web One of the principal findings of submersible-based midwater research reveals that gelatinous animals form a dominant ecological group in midwater communities. Perfectly adapted to their watery environment, their fragile construction caused them to be greatly underestimated by the trawling surveys conducted by scientists as far back as the *Challenger* expedition. Due to their low organic content, jellies can reproduce and grow quickly in response to fluctuating food supplies. Those of us who used to drag nets to sample the ocean's midwaters thought of jellies as just so much "organized water." In reality, they comprise an important but poorly appreciated portion of the food web. The jelly web is a partially closed system, because so many gelatinous animals feed on one another. Less well known is how the organic material fixed by this part of the overall web is cycled back into the rest of the deep-sea community. We do know that this material is consumed by a wide range of sea life from shrimps to salmon, and that most of the particulate matter produced by the jellies harbors populations of bacteria.

Patches of pigment in the skin of these squid (above and below) can shrink or grow, allowing the squid to change color and perhaps to signal each other.

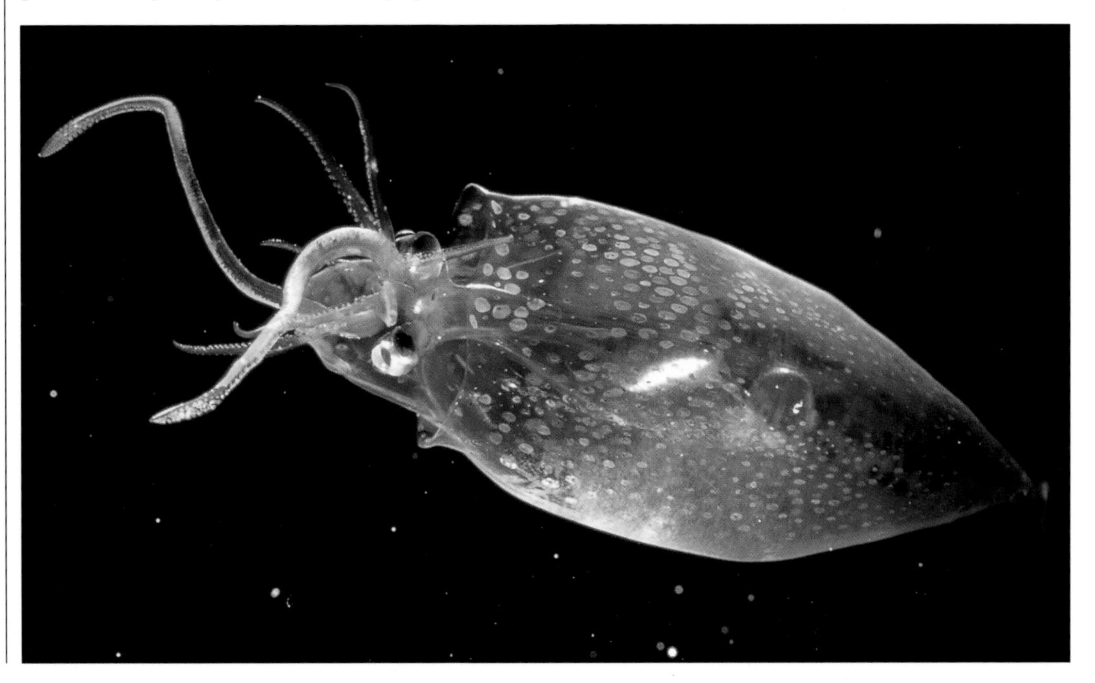

Predators in midwater come in many shapes and sizes. This medusa (left) captures prey with its tentacles. Albacore (below) use speed and strength.

The impact of jellies on the better known parts of the midwater food web appears to be substantial. Based on our preliminary measurements of abundance and feeding patterns, we estimate that at least one-third of Monterey Bay's abundant krill populations are eaten by gelatinous predators. This means that fragile, soft-bodied jellies are competing successfully against seasonal migrants like blue whales, albacore, squids and a host of other firm-bodied animals at the third level of the food web. When we sketch the outlines of the jelly web, we see nutrient energy entering in two ways. Phytoplankton-grazers like copepods and krill are fed upon by a variety of specialized gelatinous predators. Particle-feeders like salps and larvaceans are also part of the jelly web and are targeted by jelly predators with highly developed feeding strategies.

Discovering New Species

On virtually every deep dive series we make in a new location, we see new animal species—sometimes dozens of them, sometimes hundreds. The process of formally describing a new species begins with a thorough examination of several specimens, in order to completely understand their structure and range of variation. Then this information is compared with that for related species, and ideally with a side-by-side comparison of specimens. If significant differences exist (and deciding what's significant often gets us into trouble), then a detailed description of the new form is prepared and submitted for publication in a scientific journal. The manuscript is reviewed by experts in that group of animals and the response might be: "Sorry, that species was described by a Scottish parson from a specimen washed up on a beach of the Outer Hebrides in 1816." But if you have done your homework well, the description is published and you become the proud describer of a "new" species—even though they may have been swimming around in the ocean for a million years.

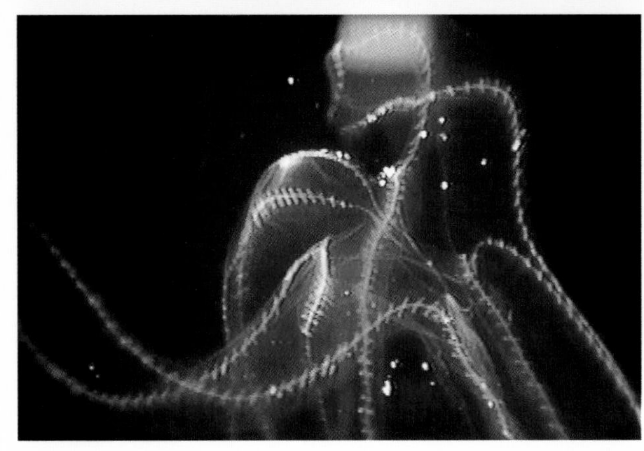

The ears of Kiyohimea usagi.

New biochemical techniques are making this process much more precise. By comparing the gene sequences of the new species with those already described, clear distinctions can be made.

Naming a new species can be fun. Scientists have named species after their sweethearts, their mothers, even their dogs. Usually though, the name is descriptive, telling us about a distinctive characteristic of the animal. A few years ago, Dr. George Matsumoto and I described a beautiful, diaphanous ctenophore based on detailed video observations from the ROV *Ventana* (fortunately, it was nearly transparent, so we could see all of its internal and external structures). We searched the scientific literature for similar species and found that in the 1940s, Japanese scientists had described a similar ctenophore called *Kiyohimea aurita*. It was clear that the two ctenophores had the same basic body form and thus belonged in the same genus—*Kiyohimea*—named after a heroine in Japanese folklore. We named the new species based on the Japanese word for what we had been calling the unknown animal, "rabbit-eared," because of two, long projections on its top that looked like a rabbit's ears. The name for rabbit in Japanese is usagi, so our new species became *Kiyohimea usagi.*

MBARI's Dr. Jim Barry has named a new species of clam from cold seeps in Monterey Bay, *Calyptogena packardana.* This is to honor the members of the Packard family, who have done so much to develop marine science, education and conservation.

If scientists were only interested in describing new species, we could keep an army of specialists happily describing away for a century. But the greater value of finding new life forms is to understand how they fit into the ecological framework of natural communities. New species are additional pieces in the puzzle of life in the ocean. We need to learn the roles they play in how the oceanic ecosystem works, so that we can live in harmony with, and conserve the ocean's living resources.

The black-eyed squid has huge eyes to let it see in dimly-lit depths.

A Whale of a Deep Diver: The Sperm Whale

Sperm whales *(Physeter catodon)* are not only stars of literature and film, like Moby Dick, they are also capable of diving to extreme depths, in excess of two kilometers. Like elephant seals (*Mirounga angustirostris*), sperm whales are air-breathing, deep-sea mammals.

The sperm whale is the largest of the toothed whales. Males grow as long as 18 meters and weigh 50 metric tons. They commonly dive one-and-a-half kilometers below the surface to feed on squids and fishes. Compared to other whales, sperm whales have an unusual, blunt-nosed appearance. They have huge heads that make up 30 percent of their body weight, with great square foreheads, wrinkled skin, small fins and slender jaws that hold up to 50 teeth. Their immense foreheads contain spermaceti oil, highly prized as light machine oil.

Masters of echolocation, which is a type of sonar, sperm whales use these reflected sound waves to find deep-sea squid. But, once they have located squids in the dark depths, how does a whale shaped like a boxcar chase down and capture jet-propelled squid? This question has been the subject of considerable conjecture and debate, but one theory suggests that a function of the spermaceti organ in the whale's big, square forehead is to focus and concentrate powerful sound waves that are emitted from the midpoint of the whale's upper lip—sound so powerful that it can paralyze a squid from a distance of at least a meter. This theory is controversial and many scientists disagree because squids appear to have no sense of hearing. Of course, that could be a defensive adaptation to the sound pulses of these whales.

Like other toothed whales, sperm whales have a complex social structure. Females and young tend to remain together in tropical and temperate waters, while the juvenile and adult males migrate as far as the polar regions to feed. The males then return to the female groups, usually to the same ones year after year, to mate.

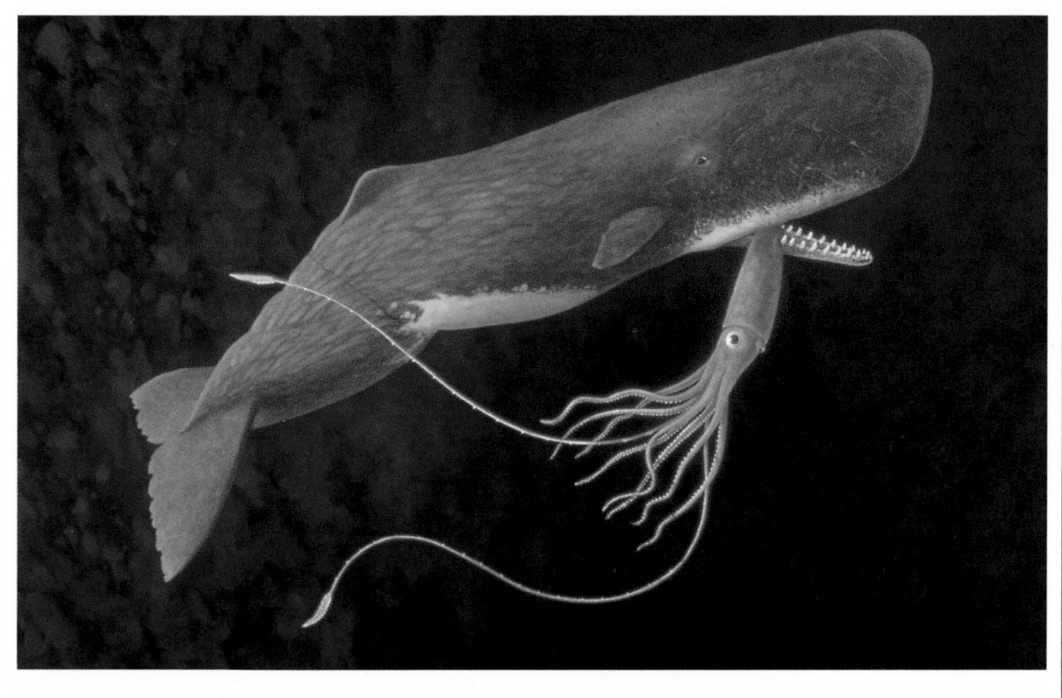

This fanciful picture depicts a struggle between a sperm whale and a giant squid (whales eat squid but not vice-versa).

Siphonophores

Siphonophores are linear colonies of individuals connected along a common stem, each with a specialized structure and role. Some individuals propel the colony; others are involved in reproduction, protection or help in food capture and digestion. Like colonies of bees or ants, they can consist of hundreds of individuals, divided into different types, with specific roles that serve the colony as a whole. Unlike the insect colonies, however, individuals in a siphonophore colony are physically connected to each other. In most species, the chain of specialized individuals issues sequentially from a cluster of swimming bells at the leading end. Siphonophores present us with a curious puzzle–whether we should think of them as an organized colony of linked individuals, or as a single, complex "superorganism." From an ecological perspective, we can regard them as living drift nets that capture fishes, shrimp, salps,

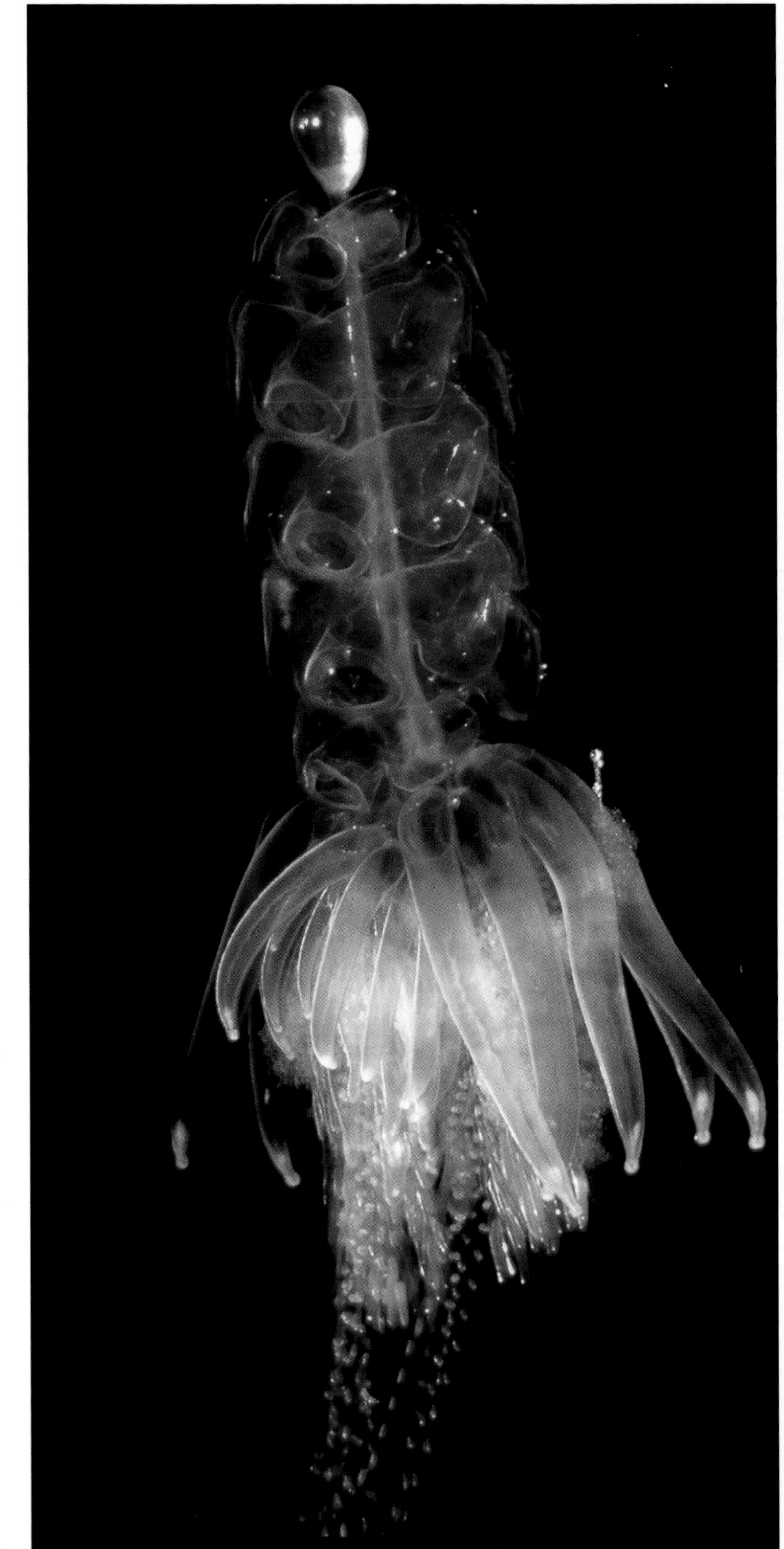

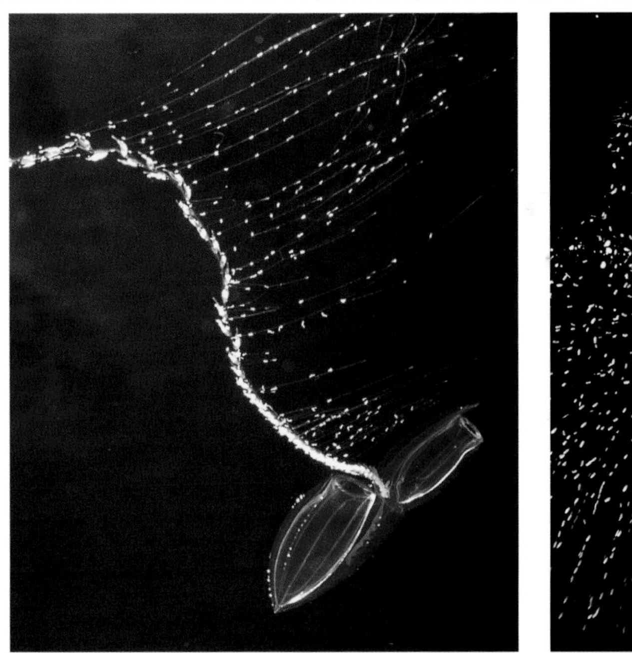

Variations on a theme: hula-skirt siphonophore (opposite), rocket-ship siphonophore (left), fireworks siphonophore (above).

ctenophores, medusae, other siphonophores and nearly anything else that swims or drifts through the water.

One siphonophore *(Praya dubia)* lives at depths between 100 and 300 meters in Monterey Bay, and with scanning sonar we have measured specimens up to 40 meters long! This makes it one of the longest creatures on Earth, longer even than the largest recorded blue whale (*Balaenoptera musculus*). *Praya* feeds on krill, copepods and other small crustaceans, as well as small fishes and jellies.

Apolemia uvaria, another large siphonophore, lives somewhat deeper at the upper margin of the oxygen-minimum layer, and grows to lengths of 15 meters or more. Both *Praya* and *Apolemia* can deploy hundreds of slender tentacles covered with stinging cells. Food captured by the tentacles is drawn back toward the stem. With the aid of finger-like palpons the prey is pushed into the open end of a nearby "stomach" or gastrozooid. When they are set out in a series, the tentacles form undulating curtains; when cast out radially, they become a cloud of slender, deadly filaments. At the beginning of the siphonophore's stem, individuals involved in propulsion usually keep the long, ropy body of the siphonophore horizontal, or swept up into an arc as it swims to regain position. The extended body acts as a sea anchor against the struggles of active prey who become tangled in its tentacle net. Specimens of *Praya* and *Apolemia* that have recently captured food have kinks in their bodies from these struggles and many swollen stomachs stuffed with their prey. Their passive feeding strategy is similar to that of human gillnetters, who seek to entangle active prey as they swim by.

The halyard siphonophore can reach 15 meters in length, with thousands of tentacles.

Occasionally long segments of the siphonophores break away from the leading end, and drift off, still alive but marooned in midwater without any way to propel themselves. Because these pieces continue to use their tentacles to capture prey, we call them "ghost tails," named after the drifting ghost nets lost in the ocean by human fishermen. The ghosts probably persist for several days at least, then break up and slowly sink.

Not all siphonophores are long. *Nanomia bijuga* is a small, five-to-25 centimeters siphonophore that feeds principally on krill, using its tentacles to capture the small, active crustaceans. It typically sits in a J-shape with tentacles splayed outward. When a euphausiid contacts and struggles against a tentacle, *Nanomia* begins swimming rapidly, which streamlines the body, with the tentacle and prey behind it. This response aligns the krill with the body, and allows *Nanomia* to play it like it was on a hook and line, reeling it in until it can be grasped by other tentacles and palpons, which then maneuver it into a nearby stomach. This behavior explains why in most individuals which have fed recently, we see the food chiefly in their posterior stomachs. After ingestion, *Nanomia* settles again into its feeding posture. But unlike its larger cousins, *Praya* and *Apolemia*, this is an impatient fisherman and it relocates frequently, deploying its tentacles in a new site every few minutes. These tactics are best suited to feeding on prey that live in clusters, and to *Nanomia's* advantage, krill often occur in patches or swarms. *Nanomia's* feeding behavior reflects a complex, well-developed colonial nervous system that belies the notion that siphonophores are merely a simple collection of specialized individuals.

The seasonal cycle of phytoplankton productivity in Monterey Bay is clearly reflected in the abundance of *Nanomia*, even though they are separated by another level of the food web. Upwelling-driven phytoplankton blooms occur during the summer, followed about a month later by large increases in grazer populations. These, in turn, lead to rapid expansion of the *Nanomia* population, which reaches its maximum just when krill numbers are at their peak. This is one important way that primary production enters the jelly web.

Most ctenophores have one of three basic body types: tongue-shaped beroids (below), diaphanous lobates (bottom, and top of opposite page), and rounded cydippids with two tentacles (bottom, opposite).

Comb Jellies Ctenophores, or comb jellies, are jellies with a wide range of body shapes and feeding strategies. Some use sticky tentacles to feed, some capture prey with fleshy lobes and others simply engulf their prey. All of them share a common means of moving around, the ctene (pronounced "teen") rows that give them their common name—comb jellies. Ctenes are fused groups of short, slender fibers, or cilia, that occur in eight rows along the length of the animal. The

ctenes move in synchronous waves that sweep down the rows, pushing water either backward or forward to propel the ctenophore, sometimes with surprising speed and agility. In the glare of an undersea vehicle, ctenes break up the light into a rainbow of rippling colors. While this is an artifact of human intervention, it remains one of the most beautiful sights we see in midwater and we always look forward to the moment when one of these graceful animals swims into our field of view.

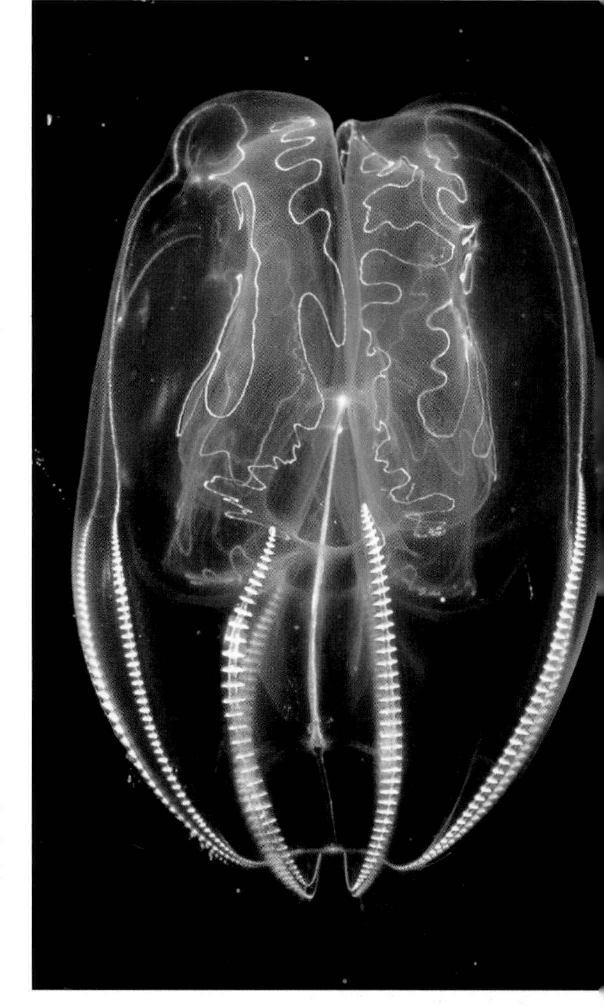

Beroë is a ctenophore that specializes in feeding on other gelatinous animals. Their bodies, shaped sort of like the sole of your shoe, range in size from five to about 15 centimeters. *Beroë's* mouth spreads wide across the front of its body, opening and closing like a keyless zipper. With no eyes and a poorly understood chemosensory capability, we don't know yet how they find their prey, but it seems unlikely (and dangerous) that they just cruise about until they bump into something good to eat. However they find it, *Beroë* usually finds small prey, like *Nanomia* and small comb jellies, and eats them whole. However, they do have hard, stubby teeth in their mouths that lets them bite off chunks of tissue from larger jellies.

Beroë itself is eaten by siphonophores and other ctenophores. Small individuals can become trapped in the labyrinths of larvacean-feeding structures. Another threat comes from amphipod crustaceans which seem to have a special taste for jellies. One day, during a *Ventana* dive to 350 meters, we observed an amphipod attacking a *Beroë*. After dropping on its back from above, the amphipod dug its hooked claws into the ctenophore's back and began to chew. The *Beroë* reacted by writhing and bucking to try and dislodge the attacker. After about 20 minutes, the ctenophore twisted its body, surrounding the amphipod, then with a convulsive flip, threw it off, now wrapped in a ball of mucus.

The rabbit-eared ctenophore (*Kiyohimea usagi*) is a lobate comb jelly that can be brought back from its habitat only as an image. It is so fragile that collecting a specimen is beyond every capture method currently at our disposal. Its body is a diaphanous pouch with two fleshy lobes suspended below its mouth and two projections like rabbit ears at the top. It uses the large lobes like soft, sticky catchers' mitts, to sweep up food. Whatever it catches is pushed toward the mouth by the lobes, two flattened projections called auricles, and tentacle branches called tentilla. Like most other ctenophores, *Kiyohimea* is completely transparent. In the dim light of the upper midwaters, transparency is an effective means of concealment from both predators and prey.

Most medusae are sit-and-wait predators who remain motionless, waiting for prey to swim into their tentacles.

Medusae The third major predatory component of the jelly web are the medusae, the animals most commonly recognized as "jellies." Their basic structure consists of a muscular swimming bell with a mouth and digestive pouch underneath. Around the margin of the bell are tentacles with stinging cells containing nematocysts, which can inject toxin or hook into the surface tissue of other animals. Radial canals within the bell carry the trunk lines of the nervous system and the gonads. Medusae are generally indiscriminate predators that feed on whatever can be caught and retained by their tentacles.

Colobonema sericeum is a small medusa with muscle bands in its bell that reflect blue-green iridescence in the lights of our undersea vehicles. Its 32 tentacles are brilliant white at the tips and deep blue along their length. Often, we see specimens with tentacles in tiers of unequal lengths. We discovered why after watching it in the wild. In normal feeding posture, *Colobonema* sits quietly with its tentacles splayed out, waiting passively to trap copepods and other small crustaceans. When startled by a predator (or an ROV), *Colobonema* darts away, rippling bioluminescence in its bell. Then the lights in the bell are doused and it zips away into the darkness, leaving bright, twisting tentacles in its wake. The number of tentacles released varies from two to ten, so the different tentacle lengths we see probably means that they are being regenerated. This process is called autotomy and it probably distracts predators in the same way as a lizard dropping its tail, but with the added deterrent of stinging cells.

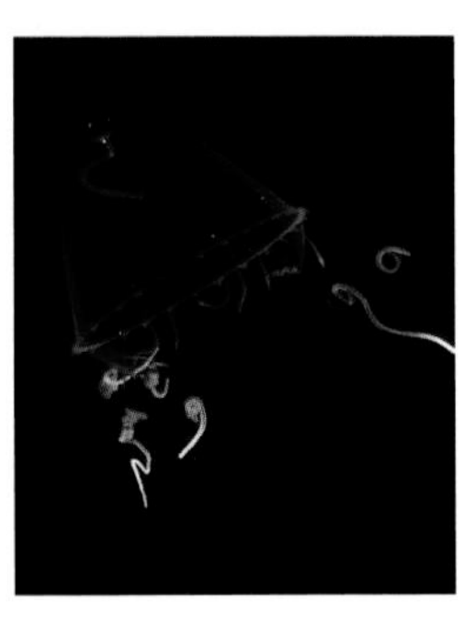

Solmissus is a relatively large medusa reaching about 10-to-12 centimeters in diameter. Unlike *Colobonema*, which relies on prey running into its tentacles, *Solmissus* is an active hunter. As it moves through the water, it holds its tentacles forward or out to the sides to increase the chances of contacting medusae and other gelatinous prey. Two species of this disk-shaped medusa occur in Monterey Bay, *Solmissus incisa* and *Solmissus marshalli*. As is often the case, when two closely related species co-occur in the same area, they live at separate depths. This keeps them from directly competing for the same food and is good evidence of ecological evolution in progress.

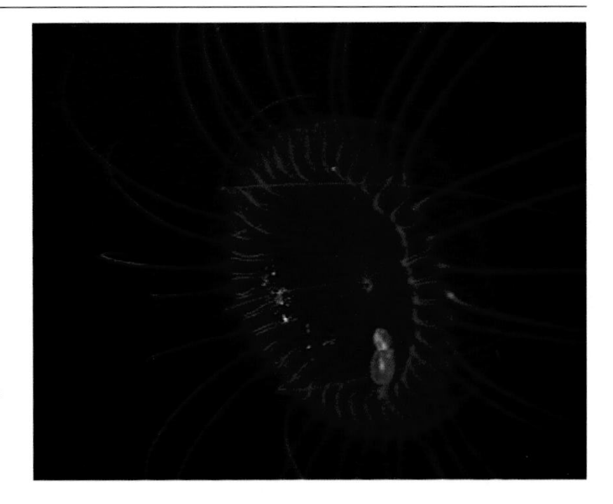

Life at Great Depths In the deep waters of the world's oceans, at depths below 800 to 1,000 meters, the midwaters contain a distinct and remarkable array of fauna, specifically adapted to life at great depths. These animals still rely on organic material generated at the surface by phytoplankton, but the distances are too great for them to move up and down to gain access to grazers. The only available light is the bioluminescence produced by the animals themselves. The absence of sunlight at these depths means that the residents rely less on vision and presumably tune in to their habitat with heightened sensitivities to sound, water movements and chemical signals. As a consequence of where they live, most of these animals are predators, relying on food that swims or sinks down to them.

Solmissus *(above) is an active medusa, hunting for prey with outstretched tentacles.*

We know far less about these animals than those that live above, because undersea vehicles are only just beginning to explore the water column at these depths. New submersible technologies will play important roles in these future discoveries. What we do know tells us that these are some of the strangest and most remarkable animals on Earth.

Anglerfish (below) entice their prey with luminous lures and visual trickery.

Anglerfish The evolutionary adaptations made by anglerfish, in order to succeed in the deep sea, illustrate the extreme environmental factors prevailing there. They also make for some pretty weird fish. Anglerfish get their name from the "fishing pole," or *illicium,* that projects from their foreheads. At the end, a bulbous light organ contains a colony of luminous bacteria which the fish provides with nutrition and shelter in return for the use of their light. Anglers are sit-and-wait predators, who use the attractiveness of their luminous lures to draw in the fishes and crustaceans they feed upon. In many cases, the light organs have complex structures with light pipes and reflective surfaces that produce a variety of luminous displays.

Most anglers live at depths below 1,000 meters, beyond the reach of even the last photons of sunlight struggling down from the surface. As a consequence, they have little use for vision and their small eyes have little ability to form images. Because their feeding strategy does not require much swimming, their bodies are neither streamlined nor muscular. They do have huge mouths and expandable stomachs, which allow them to eat very large meals, sometimes nearly as large as themselves. This suggests that meals are probably few and far between at these depths, and they must be equipped to take advantage of every reasonable opportunity to feed.

Encounters with members of the opposite sex must also be relatively rare, because some anglerfish have evolved a seemingly bizarre, but obviously effective mode of reproduction. While females can be as large as a football in some species, their males aren't much bigger than the end of your finger. Males are highly specialized, with well-developed olfactory organs and gonads, but little else. Their job is simply to find a female, presumably by following a trail of pheromones, and attach to her by biting into her flesh. Once attached, the tissues of the two animals fuse and the male degenerates to just a lump of tissue surrounding its gonads. Nourished by the female's blood the male is probably triggered chemically to release sperm as she releases eggs. Monogamy is not the rule in these relationships and I have seen females with as many as eleven males attached; the most inconveniently placed mate was lodged between her eyes. There are several variations on this theme: males that attach only briefly, and males that remain free-swimming. But the message from this remarkable reproductive pattern is that even for slow-swimming animals which are sparsely distributed, perpetuating the species must be achieved by any means possible.

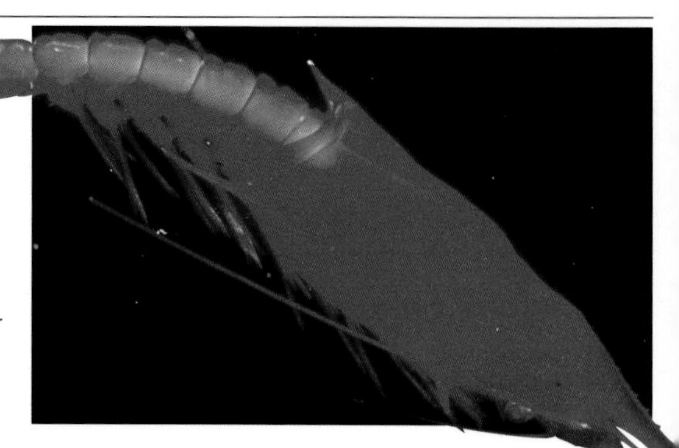

This large mysid shrimp has a heavily-armored shell, with spikes to deter predators (above). Vampyroteuthis *wraps its arms up over its head in its defensive posture (below), and uses its large eyes to collect light in deep water (opposite).*

Like many other species living so deep in the sea, anglerfish have eggs with large oil droplets that carry them to the upper layers of the ocean. The eggs hatch and their larvae develop where there is plenty of food. As the young fishes develop and grow, they work their way ever deeper until they reach the adult habitat. This process may take many months and is referred to as ontogenetic vertical migration. Deep-living animals usually grow slowly and it may take several years for a female angler to reach maturity so the cycle can begin again.

The giant red mysid (*Gnathophausia ingens)* is a deep-red crustacean with a sturdy, tank-like body that lives in and below the oxygen-minimum layer in Monterey Bay. Its incongruous red color actually serves to hide the animal, because at these depths the only natural available light is blue. We see it as red in the white lights of our undersea vehicles because its body pigments absorb all wavelengths of light except red. Under natural light conditions, the red pigments absorb the ambient blue light, no light is reflected to the eyes of a predator and the *Gnathophausia* becomes invisible. *Gnathophausia* uses another light trick to avoid being eaten. When threatened, it secretes a cloud of bright blue luminescence from pores on the underside of its body. Its swimming legs quickly spread the cloud while the animal makes its escape in the dark, surrounding water. This trick is like the dark ink clouds released by squids and octopuses in lighted waters, only in reverse.

Deep-Sea Seals: The Northern Elephant Seal

Strange as it may seem, elephant seals are deep-sea animals. Although elephant seals are air-breathing, warm-blooded marine mammals that come ashore to mate and rear their young, while at sea they spend about 90 percent of their time diving to great depths to feed, and only about 10 percent of their time on the surface. They dive to depths of 1,500 meters or deeper, and feed on deep-sea squids, sharks, rays and fishes like Pacific hake.

Living in the deep sea offers several advantages for elephant seals. When foraging at great depths, they do not compete with their shallow-water cousins for food, and while they are offshore and far below the surface they are also avoiding their main predators, great white sharks *(Carcharodon carcharias)* and killer whales *(Orcinus orca)*.

Elephant seals have some remarkable physiological adaptations for deep diving. To avoid gas-bubble disease, also called "the bends," their lungs collapse early in the dive. This prevents surplus nitrogen from entering their blood when their bodies are subjected to the tremendous pressure of depth. Like whales, their blood contains more oxygen per unit than the blood of terrestrial mammals. They also have a relatively greater volume of blood than their land-dwelling relatives.

When an elephant seal dives, its blood supplies oxygen only to the brain and other vital organs—not to the body's muscles—so the oxygen lasts longer. Their muscles already have adequate oxygen stores due to the rich supply of oxygen-bearing protein in their muscle tissues. Elephant seals also slow their metabolic rate and heartbeat during dives, making it possible to stay submerged for over an hour.

Male and female elephant seals differ dramatically. Adult males grow up to seven meters long and weigh as much as 2,700 kilograms. In comparison, females are a petite three-and-a-half meters long, and weigh in at a mere 1,000 kilograms.

A century ago, northern elephant seals had been nearly exterminated off the coasts of California and Mexico. Protective legislation has allowed them to recover.

A bright cloud in the darkness serves to startle, blind and distract a predator, while *Gnathophausia* escapes.

The vampire squid (*Vampyroteuthis infernalis*) is a living fossil of the type searched for by the scientists aboard the *Challenger* expedition. It is an archaic, present-day species from the evolutionary line that eventually split into the eight-armed octopuses and the ten-armed squids. *Vampyroteuthis* has eight arms with a beak at their center and a web between them that reaches almost to the light organs at their tips. The arms have a series of suckers and parallel lines of fingerlike cirri. Two long sensory filaments deploy one at a time from pockets between the third and fourth arms on each side. The soft, dark-brown body has fins near the tip of the mantle and a pair of large light organs that open and close like the iris on a camera lens. *Vampyroteuthis* has two large eyes that seem to shine with a beautiful opalescent blue color in the lights of our undersea vehicles.

Vampyroteuthis swims by flapping its fins and by pushing water out of its respiratory siphon. It rises only as far as the oxygen-minimum layer in Monterey Bay, which probably insulates it from the most active predators of the upper kilometer of the ocean. We don't yet know how much farther down they live, what they eat, who eats them or much about their behavior.

FOR VIDEO OR MORE DETAILS ON THESE ANIMALS:

Please visit www.mbari.org/rd

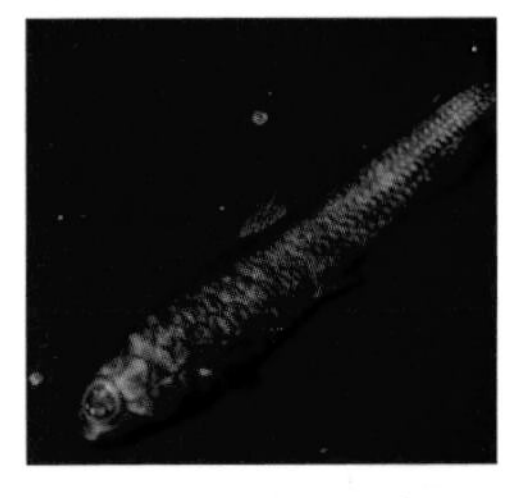

Owlfish have very large eyes, to find their prey in dimly-lit depths (above). The gulper eel (below) has small eyes and tunes into its habitat by sensing vibrations in the water.

Mysteries of the Deep Other animals inhabiting these great depths are even less well known. We have only glimpses of how they live. Owlfish are large, chunky fishes with huge, owl-like eyes and tiny mouths. They probably feed on small shrimp and jellies. They shed their large scales very easily, which we believe lets them shrug off the gripping embrace of a squid's feeding tentacle, allowing the fish to escape at the small cost of a few scales. Jellies at these depths can be enormous. One type, named *Deepstaria enigmatica* after the submersible from which it was first seen, reaches bell diameters of greater than one meter. The gulper eel (*Saccopharynx lavenbergi*) has a huge pouched mouth like a pelican, tiny eyes, a distensible stomach like an anglerfish and a long sinuous tail with a light organ at its tip, whose function we can only guess. The gulper eel is MBARI's logo animal, and represents the strange, fascinating animals we have yet to learn about as we work deeper in the sea.

4

Life on the Deep Seafloor

Life at the bottom of the deep ocean is very different from life in the fluid, three-dimensional habitat of the water above. Bottom-dwelling, or benthic animals are strongly linked to the boundary between the water and the seafloor. We humans also live at a density interface, between the air and the ground. Like us, benthic animals occur chiefly along the boundary where water and substrate meet, and most of them have vertical ranges that are very limited compared to their free-swimming relatives above. With just a few exceptions (vent and seep communities, for example), the base of the food web for seafloor animals is the same as for pelagic animals—the populations of phytoplankton near the sea surface. This means that the supply lines to the deep are very long, and that much of the organic material produced at the surface is intercepted, processed and repackaged by pelagic animals on its way down.

The Seafloor Two kinds of surfaces make up the floor of the deep sea—soft sediments and hard rocks. And there are three fundamental ways that animals inhabit the deep benthic habitat: living within the sediments or occasionally in holes bored into rock (the infauna); living on top of the sediments or rocks (epifauna); and animals which move freely through the water over the seafloor (benthopelagic fauna).

Food sinking down to the bottom either settles into the sediments or is suspended and moved along by slow currents near the bottom. The suspended material forms a layer of particles near the seafloor that is usually less than a meter thick, but can often be far more extensive. The animals that consume this material, after its long trip down, either eat the sediments or feed on the suspended particles before they settle out. As you might expect, in rocky areas, suspension-feeders outnumber sediment-feeders and in areas without rocks, it's the other way around.

In Monterey Bay, the seasonal cycles of productivity near the surface are also felt in the deepest parts of the submarine canyon, although there is considerable lag time between blooms at the surface and the resulting pulses of food to the bottom. In one important respect, the deep seafloor is a mirror image of the air-sea interface at the top of the water column—food is concentrated at the interface.

Some bottom-dwelling animals feed by burrowing through the sediments or grazing at its surface, eating the organic matter as

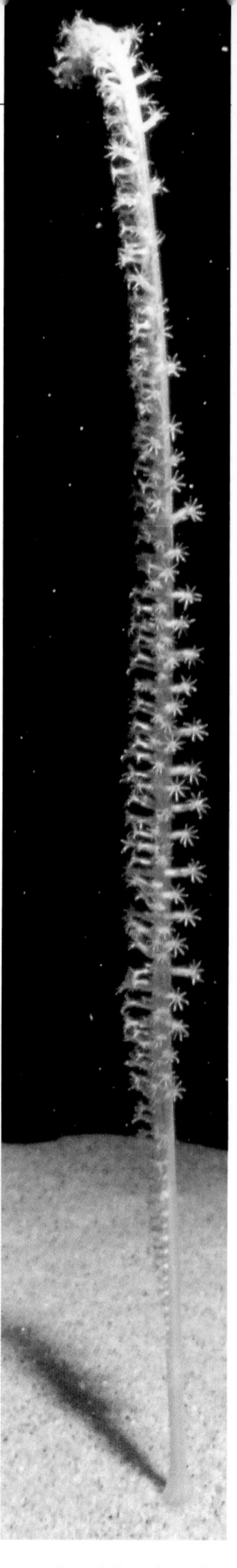

Sea whip (above) and red sea fan (opposite).

Cold Seeps

Along the axis and flanks of the Monterey Canyon, and in other deep coastal regions, scientists have found sites where water seeps slowly from the seafloor. These sites are called cold seeps to distinguish them from the hot, hydrothermal vents found elsewhere. The water that is expressed from the sediments and rocks at some of these sites is rich in hydrogen sulfide, a compound toxic to most animal life. Nevertheless, unique animal communities occur at the seeps, communities as remarkable as those around hydrothermal vents, because they do not depend on photosynthesis by plants as the base of their food chain.

Clams clustered at a cold seep.

Instead of photosynthesis driven by solar energy, the seep communities are based on a process called chemosynthesis. Energy stored in the hydrogen sulfide molecule is released by specialized bacteria, which use it to build organic compounds from dissolved inorganic carbon. At many sites, dense mats of bacteria, called *Beggiatoa,* cover the rocks where sulfide-rich water seeps through pores and cracks. In these areas, the bacterial mats are fed upon directly by animals such as galatheid crabs.

At sites where the seep water comes up through the sediments, we see another sort of community. Dr. Jim Barry of MBARI has used *Ventana* to study these communities. His research has provided fundamental new knowledge about the ecology and physiology of the unique animals that inhabit them. The dominant animals in sediment-covered Monterey Bay seep communities are clams of the genus *Calyptogena.* Chemosynthetic bacteria (actually the term is chemoautotrophic) live symbiotically within the gill tissues of the clams and provide them with their sole source of nutrition. The clams face a real challenge to keep themselves and their symbiotic bacteria alive. The clams and the bacteria need oxygen for respiration and they get it by holding the clam's respiratory siphon above the sediment interface, away from the hydrogen sulfide, and into the oxygenated sea water. At the same time, the bacteria need the hydrogen sulfide, so the clam keeps its foot in the sediments bathed by seep water and away from the oxygen. This balancing act puts half the clam in one chemical realm and half in another.

Despite the delicate balancing act, we know that the arrangement works, because some of the clams have been estimated to live for 100 years. We also know that the seep flow can be capricious. Sometimes it stops in an area and the clams die off. Sometimes the location of the seep shifts slightly or the flow rate changes. Clams must move to respond to these changes, and we see clear evidence in trails through the sediment and empty shells, that some succeed in relocating while others don't.

Other animals also inhabit the Monterey seep sites, some with symbionts and others which are only attracted to the oasis for food. Seep fluids may contain methane along with the hydrogen sulfide, which can also serve as a source of energy for some bacteria. Thus, different animal species are associated with different fluid chemistries in different locations. The more we explore the deep seafloor, the more seeps we find, and our knowledge of these remarkable communities continues to grow.

they move slowly along. Others position themselves or their feeding apparatus above the bottom and consume the suspended particles as they drift by. As much as we would like to place these animals, where they live and the ways they make a living, into distinct categories, nature seldom works that simply. Try as we might to organize them into orderly groups there are always exceptions, always contradictions and always wonderful complexities for us to try to unravel.

Club-tipped anemones (left) and spiny red sea stars (above) inhabit rocky ledges on the walls of Monterey Canyon.

Life within the Sediments Worms are among the most abundant forms of life found among deep-sea sediments. Some remain completely submerged as they tunnel through it, by eating. Others dig burrows or make tubes that open up to the surface so they can feed on the organic matter settling there, or passing by in the waters just above. Those who feed below the surface often use a large proboscis which they push forward into the sediments, then draw back into their body. Doing so carries organic matter, including bacteria, back into the gut while it pulls the worm along. Often, these worms leave spiral coils of waste in their wake, making distinctive marks on the bottom.

Tube worms have lightning-fast reflexes to pull their feeding tufts back into the tubes when threatened.

Tube-dwelling worms are less mobile and rely on tufts of arm-like extensions that reach out of the tube to collect food from the surface of the seafloor. Many species cement grains of sand to their tubes and when a predator approaches, they pull back inside for protection. Cruising above a field of these worms in a submersible triggers a wave of retractions of their feathery arms; a seafloor that looked fuzzy with feeding tufts only moments before becomes smooth and clear as they hide.

Clams, mussels and other bivalve molluscs are usually sediment-eaters or filter-feeders, although the latter become less common the deeper you go. Most frequently, they lie buried in the sediments, reaching above the bottom-water interface with the siphons they use for respiration, and their feeding appendages. They slide their muscular foot out between the two halves of their shell to slowly pull themselves through the sediments. These animals make many of the small pits, mounds and furrows that we see on the seafloor. Pits result from the scraping of their feeding appendages as they draw particles of sediment across the surface and back into their digestive system. Often these create a pleasing pattern of grooves radiating out from the center. Mounds result from processed sediments ejected upward from the excurrent siphon, only to fall back down around the site where the animal is buried. Furrows are dug from below as the bivalve pulls itself by its foot to a new location.

Life Attached to the Seafloor

This stegasaurus sponge is about the size of the brain of its dinosaur namesake (above). Particles too big to pass through a sponge's outer pores collect and provide food for grazing brittlestars (opposite).

Sponges form a diverse and important component of the deep sea epifauna, living attached to rocks as well as on soft surfaces. Some are stalked, some rest directly on the bottom and others form an encrusting layer that covers the surface. They appear in a wide variety of colors, shapes and sizes, and have been found at depths as great as 9,000 meters. These simple, primitive animals have a mode of feeding based on pumping water through a network of canals. Water enters a sponge's body through small pores on the outside, carrying food and oxygen through a series of channels to centralized chambers. Cells with hair-like flagella on the inner linings of the sponge beat back and forth independently to create the flow. Between the inner and outer skin is a skeleton of spicules, usually made of silica, formed into spiky shapes that set each species apart. Water expelled from the interior of the sponge carries away wastes, and during spawning, the eggs and sperm that will combine to form the larvae that settle and become new sponges.

With no muscles or nerves, sponge behavior is sort of an oxymoron; they don't do much besides sit there and pump water. Nevertheless, we have learned to appreciate them for the brilliant splashes of blue, red and yellow they provide as the lights of our vehicles sweep over the walls of the Monterey Canyon. Some of the vase-shaped sponges in the canyon are as large as lawn chairs, and we are tempted to land our vehicles on them for a break during the dive. Sponges provide homes and shelter for a variety of other animals, like brittle stars, fish and shrimp. The filetail catshark *(Parmaturus xaniurus)* uses sponges that project out from the submarine canyon wall as a safe repository for their eggs. Predators are relatively few, but certain fishes, sea stars and snails feed on sponges, despite their bristly texture and the toxins they produce to keep from being eaten. Some deep-living sponges, like the delicate glass sponges which seem to be intricately woven from spun glass, are exquisitely beautiful, but few of us ever get to see them.

Filetail catsharks (above) lay their eggs on sponges and rocky outcrops. The egg cases have long tendrils that hook them to the substrate.

Sea pens, sea fans, sea anemones, and soft and hard corals are relatives of the medusae and siphonophores that live up in the water column. All are cnidarians, and like their kin living in the waters above, they have stinging cells for capturing their prey. Bottom-dwelling cnidarians feed on copepods, shrimp and other small animals, as well as on organic particles carried by near-bottom currents. Sea pens and fans hold their feeding apparatus off the bottom on stalks. The pens often resemble feathers stuck in the seafloor and are named for their resemblance to old-fashioned quill pens. Sea fans have a latticelike arrangement of supports for their tentacles which gives them their name.

Both sea pens and sea fans display brilliant bioluminescence, puzzling scientists aboard the *Challenger* as well as today. What purpose do these displays serve? Dr. James Case, of the University of California at Santa Barbara, believes that the rippling waves of light surging away from any point of the body that is touched, deflect the attacks of swimming predators.

The droopy sea pen bends with the flow of bottom currents, capturing small zooplankton with its tentacles (below).

Hot, Hot, Hot Vents

In 1977, scientists studying the Galapagos rift zone with the deep submersible *Alvin* made an amazing discovery. On an otherwise barren ocean floor 2,700 meters below the surface, they discovered hydrothermal vents surrounded by a community of never-before-seen animals.

Hydrothermal vents—also called hot vents—are created when sea water trickling through cracks in the Earth's crust is heated to extraordinary temperatures, then bursts through the vents like a submarine geyser. The water in hot vents can be hotter than 350° centigrade, the temperature of molten lead. Vent water contains chemicals and sulfide minerals that precipitate when they contact cold sea water, forming chimney-like structures and, in some cases, creating black "smoke." One famous "black smoker" in the Juan de Fuca Ridge off the Oregon coast, sports a "chimney" taller than a 13-story building, earning it the name "Godzilla."

Not only are hot vents physically spectacular, so is the life that flourishes around them. Hot vents are undersea oases on the deep seafloor, nourishing very unusual forms of life that thrive in the energy-rich environment. Typically, animals that surround hot vents include giant tube worms up to one meter long, clusters of clams as much as 32 centimeters across and a variety of mussels, crabs, shrimps and fishes. But, the most remarkable aspect of hot-vent life is that it thrives on bacterial chemosynthesis—the synthesis of food using chemical energy — instead of photosynthesis based on light.

A black smoker in full bloom (top). Red-tipped, giant tube worms (bottom).

Sea anemones, similar to those we see in tide pools, range to depths beyond 10,000 meters. Some species attach to rocks, some root in the sediments, some burrow into the bottom and build tubes lined with cemented grains of sand. Still others gain mobility by attaching to the hard shells of crabs and snails.

Soft corals come in a wide range of shapes, sizes and bright colors, even where no natural light penetrates to show them off. The mushroom soft coral *(Anthomastus ritteri)* is a common soft coral on the walls of the Monterey Canyon. Its short, thick stalk supports a bulbous top with tentacle-bearing polyps. From a distance, it looks like a pink mushroom with red palm trees growing out of the top. Crabs and fish prey on both the soft corals and sea anemones. When attacked, they contract their soft tentacles and wait until the predator moves on.

Bottom-dwelling tunicates live very different lifestyles than their relatives, the

The pom pom anemone (left) and the mushroom soft coral (above and bottom) are cnidarians, relatives of the midwater medusae and siphonophores. All share the ability to capture prey with stinging cells on their tentacles.

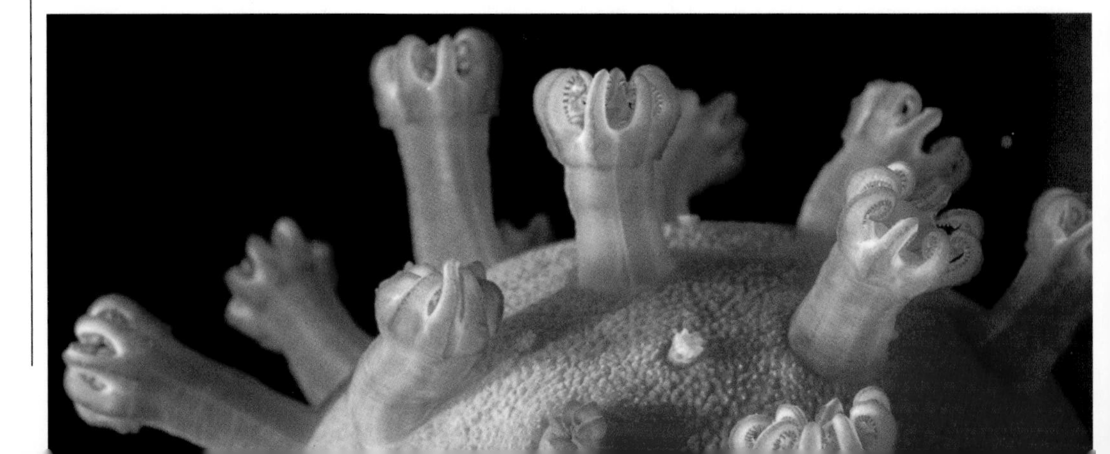

salps and larvaceans. Because they are attached to the seafloor, their bodies must be much more rugged and they lack the ability to move from food-poor areas to "greener pastures." But both the pelagic and benthic forms filter-feed using mucus to separate the particles they eat from the water that carries them. Most deep benthic tunicates have a globular shape, with a siphon for water to enter and a separate siphon for it to leave. In between the siphons are sieve-like structures covered with tiny filaments called cilia. Mucus, secreted within the body, is moved across the sieve by the cilia, which also draw water and particles into the siphon. The mucus traps the particles, carrying it into the tunicate's stomach, while the filtered water passes out through the other siphon.

On the walls of Monterey Canyon lives an unusual, carnivorous tunicate named *Megalodicopia hians*. The currents that sweep along the walls of the submarine canyon often carry copepods, krill and other zooplankton. Many invertebrates which live attached to the seafloor, like anemones and soft corals, feed on this rich source of food. *Megalodicopia* has modified the basic tunicate anatomy to take advantage of this feeding bonus. The siphon carrying water into the tunicate has been enlarged, and its shape and musculature has changed to allow it to open and close like a set of jaws. The first time we saw *Megalodicopia* with a live krill inside was during a *Ventana* dive on the south wall of the submarine canyon. At first, we were skeptical about how the krill had gotten inside (everyone knows that tunicates are filter-feeders) but soon we watched and recorded how the little carnivore snapped its "mouth" down whenever a euphausiid swam close enough to catch.

Most tunicates, whether they are salps in mid-water or sea squirts on the bottom, are filter feeders.

The exception to the rule of tunicate feeding is this predatory species that lives on the walls of Monterey Canyon.

Most sea cucumbers live on the bottom (above), munching away on the organic particles that accumulate after their long fall from the ocean's upper layers. A few cukes have moved up into the water column (below).

Moving about on the Seafloor Sea cucumbers, or holothurians, are among the most commonly observed animals of the deep seafloor. They move along the bottom on multiple "tube feet," using a ring of tentacles around their mouths to collect and ingest the upper layer of sediment. Most have long bodies, but some are rounded and look like plump little piggy banks on stubby legs. Occasionally, we see large groups of holothurians gathered around food, and in the 1960s, scientists aboard *Trieste* observed a large herd of them marching along the bottom toward food upstream.

Sea cucumbers come in many shapes and find many ways to make a living. Some species burrow into the seafloor to gain protection from predators as they slowly eat their way through the sediments. Most species walk upon the surface, have tough, heavy armor-like skin, or toxins in their skin to keep from being eaten. A few have become mobile by eliminating heavy body parts, developing webbed swimming structures at the front and back and by spending most of the time up in the water to avoid benthic predators.

One of these swimming sea cucumbers reveals how some deep-sea animals use bioluminescence as a defense. The gelatinous cucumber *(Enypniastes eximia)* is bulb-shaped, about the size of a softball. Its body is covered with a thin layer of sticky red skin that is easily shed and contains hundreds of tiny light-producing granules. After spending time on the bottom using its tentacles to pack its gut with food, *Enypniastes* lifts off and swims upward. Once away from the threat of benthic predators, it drifts slowly along, digesting its food until it is time to drop briefly back to the bottom for another meal. While it isn't tough or toxic, *Enypniastes* is well-protected by its skin. When a predator strikes, the skin lights up and quickly peels away. Because it is sticky, it adheres to the predator, making the attacker glow and thus become vulnerable to other predators nearby. After the attack, the cucumber quickly grows new

This swimming sea cucumber, Enypniastes, *uses light produced in its skin to foil predators (above).*

skin. People use similar tactics to foil bank robbers. Exploding dye packets in sacks of stolen money mark the robbers with a purple stain, making it easy for the police to find them.

Brittle stars, or ophiuroids, are small relatives of the sea stars. Like the sea cucumbers, they are echinoderms. They have five long, slender arms attached to a central disk containing the mouth and internal organs. Scales or spines lining the arms give them a furry appearance. Unlike sea stars, which move about on their tube feet, brittle stars use their arms for locomotion. One or two arms reach out in the direction it wants to move, and pull the body forward while the other arms trail along. Changing direction is simply a matter of pulling with a different arm. In many parts of the deep sea, brittle stars are the most abundant animals visible on the bottom, sometimes numbering in the millions over a small area. They occur worldwide and have been found at depths as great as 7,000 meters.

Brittle star (above) and basket stars climbing a sea pen (below).

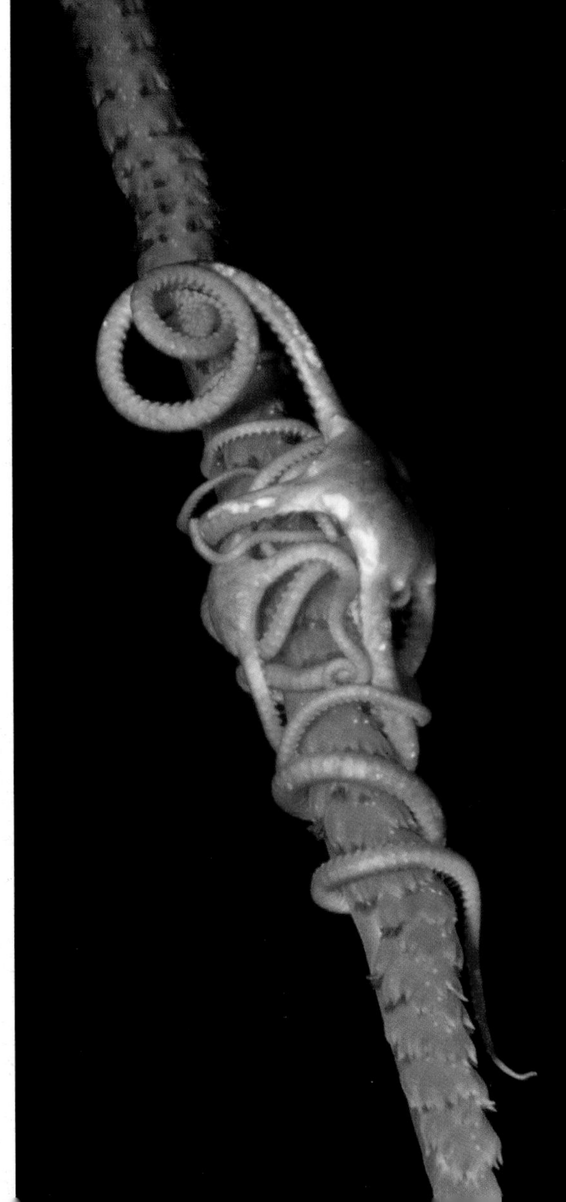

Brittle stars feed on small organic particles they sweep from the surface of the seafloor or collect from moving water above them. They also feed on living prey and can be surprisingly fast when they reach up into the current to snag a copepod or tiny fish (at least the fish are surprised). Another curious trait of brittle stars is that they like to climb. We often see them several centimeters above the bottom, with two or three arms wrapped around the stalk of a sea pen or stalked sponge, reaching out into the current with the other arms to collect food. Sometimes it seems that every sea pen in sight has one or two brittle stars that have shinnied up out of the mud to get a clearer shot at the current-borne food.

A brittle star's arms break off easily, earning them their common name. Fortunately for the brittle star, they grow back readily. This is an important advantage because it allows them to sacrifice an arm or two to a predator who grabs them, while the main part of the body and the remaining arms escape (another case of autotomy).

Deep-living benthic crustaceans and their arthropod kin are typically carnivores and scavengers, although a few are known to be sediment-feeders. One of the oddest-looking is the pycnogonid sea spider (*Pycnogondia helianthoides*). These animals have extremely long, arched legs, sometimes more than ten times the length of the body. The long legs probably spread the weight of the animal out over a wide area as it walks over soft sediments. The legs can be opened and closed like an umbrella, to lift the sea spider up off the bottom.

A scavenger amphipod from the bottom of the Mariana Trench.

Amphipods are the dominant scavengers of the deep trenches, and they are usually the first to arrive at a food fall or a baited trap placed by scientists. Most amphipods are good swimmers when food is present and they travel up-current along a gradient of chemical cues to find it. Smaller species work close to the bottom where currents are somewhat slower and they are more successful at exploiting smaller food falls. The larger species work higher off the bottom, in faster currents and are better adapted to take advantage of larger food falls with significant odor plumes.

Squat lobsters *(Munida quadrispina)*, also called galatheid crabs, can be found in cracks and holes, on ledges and ridges, nearly the whole depth range of the Monterey Canyon. Their strong, muscular tails curl under the rear of their bodies, and when threatened or startled, the squat lobsters flip their tails repeatedly in a very effective and speedy escape response. Galatheids not only scavenge food, but also steal it. We have often watched them from *Ventana*, stealing food already captured by anemones. Some squat lobsters appear to make their homes next to tube-dwelling anemones, thus gaining protection as well as a place to swipe a snack.

Squat lobsters (above) are really crabs, not lobsters, and are close relatives of the hermit crabs (below).

Hermit crabs also inhabit great depths, as do crabs that look just like those that went into your last crabmeat cocktail. Some deep-living crabs are fearlessly aggressive. When startled by an undersea vehicle, they turn to face it, rocking back on their rear legs, and raising their two largest claws in a defiant posture to their two-ton opponent that says, "Are you sure you want to mess with me?"

Snails, or gastropod molluscs, are found all the way from the intertidal zone to the bottoms of deep trenches. Some species, usually those with coiled shells, feed on worms or bivalve molluscs. The worm-eaters often kill their prey with a poison secreted by their salivary glands. Those that prey on other molluscs use their radula, a conveyor belt-like tongue with hundreds of sharp, hard teeth, to drill holes through the shells of their prey.

Snails with uncoiled, limpet-like shells are usually scavengers or grazers on the sediments. Some of the scavengers feed on unexpected sources of nutrition, like squid beaks, wood, and whale bones, as well as shark egg cases and the carcasses of crustaceans.

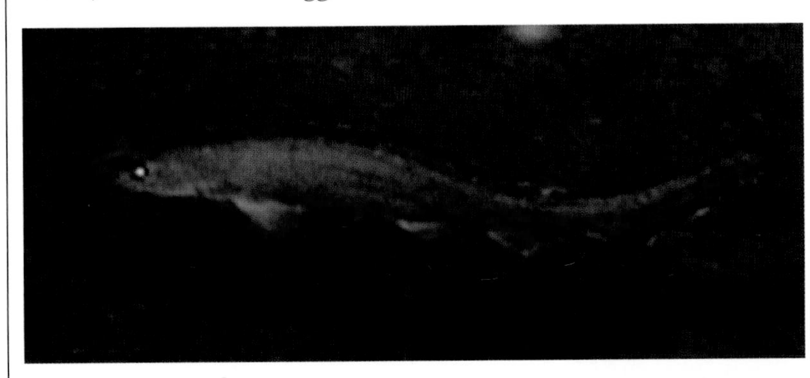

Catsharks occur most commonly near the bottom, in continental slope habitats. They also forage up into the water column, sometimes hundreds of meters from the seafloor.

Fishes A wide variety of fishes live on or near the floor of the deep sea. Some are wide-ranging predators that cruise above the bottom in search of prey. Others are scavengers that seek out the carcasses of dead animals, many of which sink down from the waters above. These mobile fishes cover a lot of territory in their roaming quest for food, and in order to reduce the energy costs of constantly being in motion, their bodies are kept at a state of nearly neutral buoyancy. They can achieve this in several ways.

Sharks, rays and skates (elasmobranchs, or fishes with skeletons made of cartilage) usually have livers filled with oils that are lighter than water. Other fishes (some of the teleosts or bony fishes) have gas-filled swimbladders to provide buoyancy. When fishermen pull these fishes up quickly from the depths, the gas expands, pushing their internal organs outward and giving them a bug-eyed appearance.

Another way to reduce body weight is to decrease the amount of calcium in bones and increase the ratio of water to protein in muscles. While these modifications can make fishes seem flabby when we see them at the surface, at depth their lightweight tissues function very well.

In contrast to the mobile fishes, another group makes their living as sit-and-wait predators. These species usually have heavy bodies with large mouths. They rest on the bottom waiting for their prey to approach. This is an energy-conserving feeding strategy and these fishes typically have very low metabolic rates.

While bioluminescence is widespread among the animals that-inhabit the water column, it is much less common for those living on the bottom. As a result, many fishes that live below the deepest penetration of sunlight have very small eyes. Instead of vision, they rely on other senses to find food. One is the ability to detect chemical cues in the water, sort of a combination of our senses of smell and taste. Another is the ability to detect low-frequency vibrations created by the movements of other animals. They can do this by using a series of pores, along their sides and snout, called the lateral line. Behind the pores are sensory filaments set in a jellylike matrix. Slight movements of water caused by other animals create displacements of the filaments, triggering signals in the nervous system. The arrangement of lateral line pores allows the fish to tell the approximate size, direction, distance and speed of animals swimming nearby. Some sharks have an even more exotic sensory capability that allows them to detect the electrical fields generated by other animals.

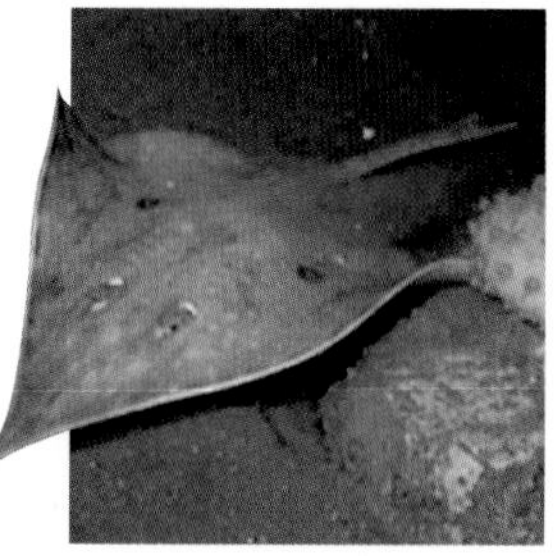

Skates seldom travel far from the bottom and the protection they gain by blending into it.

Among the most abundant bottom-dwelling, deep-sea fishes are rattails (Macrouridae). Common along continental shelves, they live as deep as the abyssal plains. Rattails have large heads and the long tapering tails that give them their common name. Elaborate, extensive lateral line systems cover their heads and bodies. They range in size from 20 centimeters to about a meter and a half in length. We usually see them facing into the current just a few centimeters above the bottom, with their snouts angled downward, slowly sculling forward with their elevated tails. Some species feed on animals swimming just above the bottom, while others pick their food up off the seafloor. Still others use pointed, hardened snouts and underslung mouths to root in the seafloor for worms and other small invertebrates. They thrust their snouts into the sediments, take in a mouthful and push the muddy slurry back out through their gills. Small projections on the inner arch of the gills act like the teeth of a comb and trap the food while the slurry passes through. The species that root in the sediments often have "barbels," filaments covered with taste buds that dangle beneath their chins. They use the barbels to probe the sediments for food, sort of like dragging your tongue through the mud.

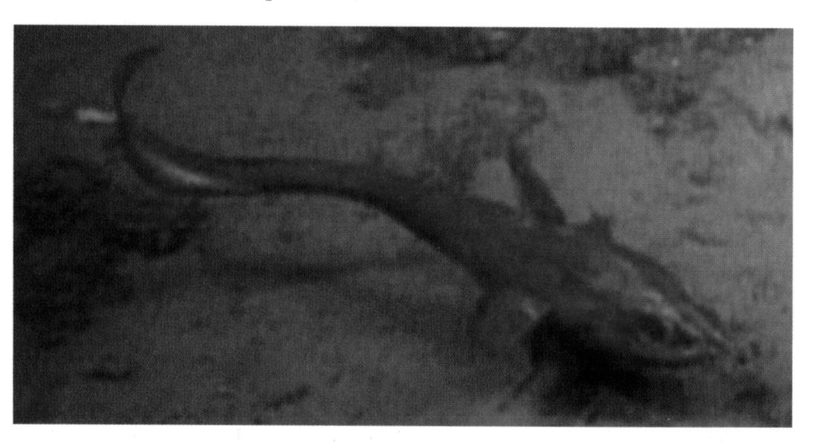

Rattails are probably the dominant fishes in deep waters of Earth's continental shelves. They range from the Arctic to the Antarctic and occur in all oceans.

Many rattail species display complex light organs near their anus. The structures contain lenses, mirrored surfaces and blackout sections. Both males and females have the light organs so they are apparently not used for finding mates, and they are inappropriate for counter-illumination to protect from predators. We don't yet know their function. We do know that they contain colonies of luminous bacteria. Portuguese fishermen sometimes squeeze the bacteria out of these organs and wipe their bait with the glowing fluid to create luminous lures that last for hours. Several species of rattail also produce sound by drumming the muscles along the sides of their swimbladders. Again, we don't know how or why they use the sound.

Rattails tend to live in small groups, although each individual usually patrols its own territory. When one fish finds food, it abandons the boundaries of its patrolled territory while it feeds, and its neighbors know to move in that direction to find food. Then their neighbors find unpatrolled borders and also begin to move in the direction of the food. Ultimately, a large group of rattails will converge on a big food source. This is called the collapsing territories process and it is also found among wolves, vultures and other terrestrial scavengers.

FOR VIDEO OR MORE DETAILS ON THESE ANIMALS:

Please visit www.mbari.org/rd

In recent years, rattails of the genus *Nezumia* have become an important commercial fishery stock in Monterey Bay. They have a mildly flavored meat that is low in oil content and they are sold in the market as "Grenadier."

Other deep-sea fishes include the tripod fish, sit-and-wait predators that take advantage of the currents carrying small animals along, just above the seafloor. They get their name from the long, rigid extensions of their pectoral and lower tail fins. These long fins allow the tripod fish to stand up off the bottom, resting on the three supports. Like the brittle stars that climb up the stalks of sea pens and sponges, they gain a real feeding advantage by placing themselves a little above the bottom. We usually see them facing into the current, often at the crest of a mound or rise. They feed on copepods and small shrimp, using their lateral-line sensors and their pectoral fins, which they hold upward and outward from their bodies, to detect the presence of approaching prey.

A tripod fish faces into the current with its sensitive pectoral fins spread wide.

Biodiversity The walls and floor of the Monterey Canyon are places of enormous biological diversity, like the deep seafloor worldwide. Human exploration of these areas has only just begun to reveal the magnitude of their richness in species. While we can categorize them in broad ecological terms and place them into larger taxonomic groups, the number of species yet to be discovered is as challenging as counting all the stars in the sky.

The advantage of protecting and maintaining deep-sea biodiversity includes far more than just food for human consumption. Natural products hold tremendous potential for human benefit in pharmacology and biotechnology. The toxins produced by sponges and sea cucumbers are being studied as antitumor agents, and several show great promise as future medicines. Likewise, the enzymes found within some hydrothermal vent species function at high temperatures and pressures. These traits have value for industrial solvents (many of which are enzymes) which must function under similarly harsh conditions.

A rare benthic medusa.

5

People and the Deep

Historically, the deep realm has held our attention as a source of food and treasure, mystery and exploration, and as a dumpsite of limitless dimensions. Far from remaining a pristine environment, the deep sea holds the debris of our civilizations. In many cases, particularly during the past century, we have ignored the results of dumping all kinds of refuse into the oceans—from household garbage to barrels of radioactive waste. Flying above the deep seafloor in a submersible often reveals the refuse of our lives, hidden from eyes at the surface. Off port cities like San Francisco and San Diego we see huge mounds of junk: cars, trash of all kinds, even kitchen sinks. While the seafloor beneath the open ocean is less impacted, it is not unusual to find beer cans and plastic bags of garbage.

The coast of California renders a prime example. After World War II, the Atomic Energy Commission disposed of low-level radioactive waste by embedding it in cement poured into 55-gallon steel drums. A disposal site was designated near the edge of the continental shelf in the Gulf of the Farallones, off San Francisco. The disposal methods were less than precise—who could imagine then that they would ever pose a problem? Thousands of drums (over 47,000 of them) dropped from ships in the general disposal area were scattered haphazardly about on the continental shelf or rolled down the continental slope to deeper sites. Dozens of years later, most of those steel drums have corroded away in the sea, exposing the cement and its contents.

Seawater corrodes the steel drums, degrades the cement and eventually exposes the low-level radioactive waste material they were intended to protect.

At higher levels of radioactivity, thousands of metric tons of spent fuel from nuclear power plants and decommissioned weapons have been accumulating on land for half a century. Some nuclear engineers and scientists propose the deep seafloor as a nuclear graveyard for the high-level waste, using boreholes drilled into the abyssal muds to bury waste—out-of-sight of us on land. This is a frightening prospect, because if we find that it wasn't such a good idea after all, there is no way to get this material back.

Pollution For the most part, our impacts on the ocean originate on the continents. More than half of the people in the United States live along the coast, where they fish and farm, labor and play, and influence lives under water. Land laid bare of vegetation for housing developments or agriculture becomes a prime target for erosion. Rain-swollen streams and rivers drag loose sediments with their outflow. In the ocean, the runoff obscures the sunlight needed by seaweeds and phytoplankton for photosynthesis with negative affects on the whole food web. Both natural minerals and manufactured chemicals wash into the sea with the suspended sediments. Runoff from farmlands transports herbicides and pesticides including organochlorides into the ocean, where they are taken up by phytoplankton and particles. Thus they enter the food web and because they are not biodegradable, they accumulate in the fatty tissues of higher organisms as a toxic threat.

A dover sole parks beneath a rock ledge covered with the stacked egg cases of a deep-sea whelk. While tough cases protect the eggs from predators, they are vulnerable to chemical pollutants.

The billions of tons of particles that pour off the continents each year also play a role in ocean chemistry. Some of the particles settle out virtually unchanged from their form on land; others dissolve into chemicals that recombine with compounds in sea water to form new solids. Heavy metals such as mercury, lead and cadmium can become concentrated regionally. Mixing patterns in the oceans can carry them broadly, in concentrations that do not occur naturally. We don't know what the effects may be.

Deep-Sea Fisheries The overharvesting of nearshore waters has devastated a number of fisheries. Atlantic cod, haddock and flounder fisheries have collapsed. Pacific squid and salmon are threatened, and abalone are near extinction. In the coastal waters of the United States, the status of many fisheries remain a mystery, with only about a third of over 700 major marine species well understood. With the scarcity of familiar fishes in shallow waters, fishing is driven ever deeper. New techniques for locating and capturing ever-larger numbers of fishes add pressure on the less-familiar fisheries from deeper habitats. Their biology and natural cycles of abundance aren't well understood. What we've learned about their lives just complicates the harvesting questions. In the depths of Monterey Bay, for example, scientists have found beds of slow-growing, deep-sea clams that may be 50 years old or more. These long-lived clams don't spawn each year, but at irregular intervals, perhaps under specific conditions that we don't yet understand. Similarly, many deep-sea fishes, like the thornyhead and deep rockfishes are long-lived

Orange roughy inhabit seamounts beneath the open ocean. A lack of knowledge about their life history has led to their being nearly wiped out.

species, slow to grow and slow maturing. An individual fish may take 30 years to grow big enough to produce eggs. That big granddaddy of a fish in the market may in fact be just in its prime of life. That smaller female, caught before it reaches maturity, will never have a chance to reproduce and help maintain a sustainable population.

As fish stocks on the continental shelves decline, fishing moves into deeper water.

The deep sea seems so removed from our daily lives. Why should we care? We care because the sea holds the vestiges of the origin of species, the biological and geological history of our planet, as well as its future, and our greatest wilderness. Beyond the purely practical, the mysteries of the deep sea provide the challenge of discovery and exploration. There's so much more to know, to understand. Even in the absence of knowledge, we are changing the oceans, and the life that calls it home, at an ever-increasing pace. The mysteries lie there, deep and dark, awaiting our discoveries and stewardship.

Deep rockfish (above), spiny red star (below).

Conservation

How do we protect something as vast as the deep sea, with its diverse habitats and millions of species? The better we know and understand it, of course, the better we can preserve it. This means continued exploration and research so that we can learn how it is structured and how it works as a dynamic system.

We must carry over to the deep sea, the principles of conservation we learn from terrestrial and shallow-water experiences. Many of these lessons have been learned the hard way, but as conservators of the deep, we are responsible for ensuring that we don't make the same kinds of mistakes again.

Continued exploitation of the ocean's resources is inevitable. And while it seems simplistic, the guidelines for deep-sea stewardship are indeed simple. Nonrenewable resources should be extracted in such a way that the living resources are not threatened. The living resources should be harvested in ways that sustain basic stocks of all species. This is no more than smart economics.

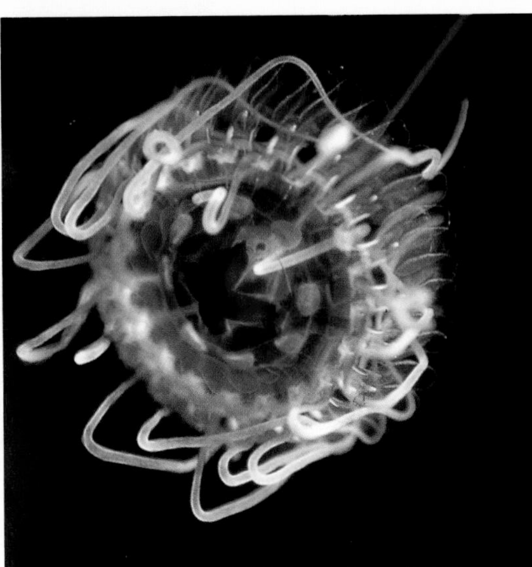

Deep-living scyphomedusa, Atolla *(above).*
A tanner crab encounters an artifact from a remote civilization. Our encroachment into the deep-sea habitat is expanding rapidly (below).

Deep-sea Animals of Monterey Bay

COMMON NAME	SCIENTIFIC NAME
SPONGES	
Glass sponge	*Acanthascus platei*
Encrusting sponge	*Acarnus erithacus*
CNIDARIANS	
Medusae	
ncn*	*Mitrocoma cellularia*
Dinner plate medusa	*Solmissus marshalli*
Golf tee medusa	*Aegina citrea*
Silky medusa	*Colobonema sericeum*
Basketball medusa	*Poralia* sp.
ncn	*Deepstaria enigmatica*
Siphonophores	
Rocketship siphonophore	*Lensia* sp.
Sesquipedalian siphonophore	*Praya dubia*
Halyard siphonophore	*Apolemia uvaria*
ncn	*Nanomia bijuga*
Soft Corals	
Mushroom soft coral	*Anthomastus ritteri*
Anemones	
Pom-pom anemone	*Liponema brevicornis*
Pennatulids	
Sea pen	*Ptilosarcus gurneyi*
Droopy sea pen	*Umbellula lyndahli*
CTENOPHORES	
Rabbit ear comb jelly	*Kiyohimea usagi*
ncn	*Bolinopsis infundibulum*
Sea gooseberry	*Pleurobrachia bachei*
ncn	*Beroe abyssicola*
MOLLUSCS	
Bivalves	
Cold-seep clam	*Calyptogena packardana*
Deep-sea clam	*Nuculana leonina*
Teaser clam	*Yoldia seminuda*
Deep-sea mussel	*Modiolus neglectus*
Gastropods	
Elephant snail	*Neptunea amianta*
Heteropods	
Sea elephant	*Carinaria cristata*
Pteropods	
ncn	*Clione limacina*
Cephalopods	
Vampire squid	*Vampyroteuthis infernalis*
Squids	
ncn	*Chiroteuthis calyx*
Cockatoo squid	*Galiteuthis phyllura*
Cock-eyed squid	*Histioteuthis heteropsis*
Octopuses	
Transparent octopus	*Japetella diaphana*
ANNELIDS	
Polychaetes	
Butt worm	*Poeobius meseres*
Tomopterid	*Tomopteris pacifica*
POGONOPHORANS	
Bearded tube worm	*Lamellibrachia barhami*

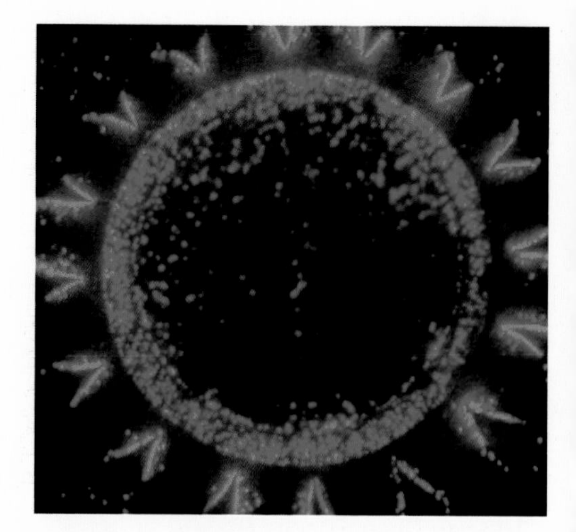

Bioluminescing medusa

Squat lobster

* no common name

ARTHROPODS	
Crustaceans	
California king crab	*Paralithodes califoriensis*
Squat lobster	*Munida quadrispina*
ncn	*Cystosoma fabricii*
Krill	*Euphausia pacifica*
Sergestid shrimp	*Sergestes similis*
Giant red mysid	*Gnathophausia ingens*
Black prince copepod	*Gaussia princeps*
Scavenging amphipod	*Orchomene decipiens*
CHAETOGNATHS	
Arrow worm	*Pseudosagitta* sp.
ECHINODERMS	
Basket star	*Gorgonocephalus eucnemis*
Deep sea star	*Hippasteria spinosa*
Multi-armed star	*Rathbunaster californicus*
Brittle star	*Ophiomusium lymani*
Pink sea urchin	*Allocentrotus fragilis*
Piggybank holothurian	*Scotoplanes clarki*
Glowing sea cucumber	*Pannychia moseleyi*
CHORDATES	
Tunicates	
Predatory tunicate	*Megalodicopia hians*
Larvaceans	
Giant larvacean	*Bathochordaeus charon*
Hammerhead larvacean	*Oikopleura villafrancae*
Redhead larvacean	*Mesochordaeus erythrocephalus*
Salps	
Spiral salp	*Cyclosalpa bakeri*
ncn	*Thalia democratica*
Fishes	
Cartilaginous fishes	
Sleeper shark	*Somniosus pacificus*
Filetail catshark	*Parmaturus xaniurus*
Blue shark	*Prionace glauca*
Longnose skate	*Raja rhina*
Sandpaper skate	*Bathyraja kincaidii*
Spotted ratfish	*Hydrolagus colliei*
Bony fishes	
Slender snipe eel	*Nemichthys scolopaceus*
Gulper eel	*Saccopharynx lavenbergi*
Owlfish	*Bathylagus milleri*
Bigtooth bristlemouth	*Cyclothone acclinidens*
Silver hatchetfish	*Argyropelecus lychnus*
Pacific viperfish	*Chauliodus macouni*
Dragonfish	*Tactostoma macropus*
Blackdragon	*Idiacanthus antrostomus*
Barracudina	*Lestidiops ringens*
Northern lampfish	*Stenobrachius leucopsarus*
Pacific hake	*Merluccius productus*
Flatnose mora	*Antimora microlepis*
Pacific grenadier	*Coryphaenoides acrolepis*
Pallid eelpout	*Lycodapus mandibularis*
Opah	*Lampris guttatus*
King of the Salmon	*Trachipterus altivelis*
Fangtooth	*Anoplogaster cornuta*
Shortspine thornyhead	*Sebastolobus alascanus*
Sablefish	*Anoplopoma fimbria*
Medusafish	*Icichthys lockingtoni*
Blacktail snailfish	*Careproctus melanurus*
Deep sea sole	*Embassichthys bathybius*
Pacific dreamer	*Oneirodes acanthias*
Blackdevil anglerfish	*Melanocetus johnsonii*

Tanner crab

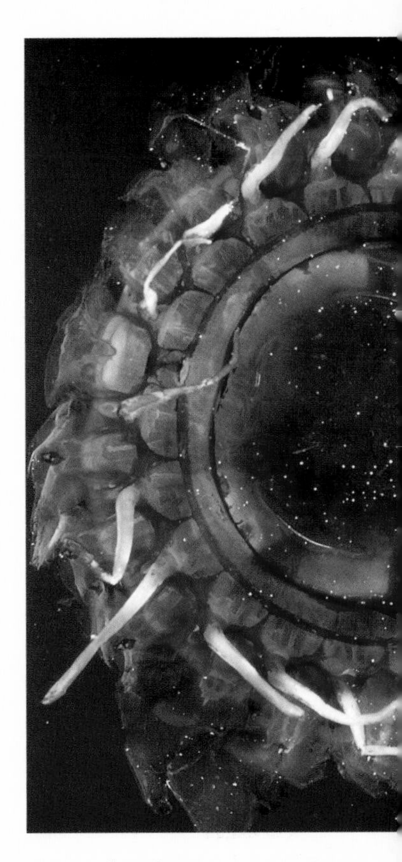

Deep red medusa

INDEX

abyssal 13
Alvin 11, 12, 18, 20, 22, 65
amphipod 51, 70
anemones 64, 66, 67
anglerfish 15, 53
Anoplogaster cornuta 10
anoxia 10
Anthomastus ritteri 67
Apolemia 49, 50
aquifers 27
Architeuthis 5
Arctic 17
Argyropelecus lychnus 38
Atlantic 17
Atomic Energy Commission 74

bacteria 15, 53, 60
Balaenoptera musculus 49
barracudina 18, 40
Barry, Jim 45, 60
Barton, Otis 16, 19
Bathochordaeus charon 37
Bathylagus milleri 9
Bathypteroidae 17
bathyscaphe 17, 18, 25
bathysphere 6, 16, 17, 19
Beebe, William 6, 16, 17, 18, 19
Beggiatoa 60
Bermuda 16, 19
Beroë 34, 50, 51
bioluminescence 16, 38, 52, 53, 64, 68, 72
biotechnology 73
black smokers 65
Bostelmann, Else 17, 19
brittle star 59, 63, 69

California Current 28
Calyptogena packardana 45, 60
Carcharodon carcharias 56
Carmel Canyon 27
Case, James 64
catsharks 63, 64, 71
Caulophryne 15
Ceratiodea 15
Challenger Deep 18
chemistry 12
chemoautotrophic 60
chemosynthesis 60, 65
Chiroteuthis calyx 41
clams 28, 45, 60, 63
Clelia 13
cnidarians 64
cold seeps 12, 13, 60
Colobonema 31, 52
colonial siphonophore
comb jelly 50, 51
copepods 10, 36, 38, 40, 49, 67
coral, hard 64
coral, mushroom soft 66
coral, soft 10, 64, 67
crabs 14, 67, 70
ctenophore 45, 49, 50, 51
ctenophore, rabbit-eared 45, 51
cucumber, gelatinous 68
Cyclosalpa 37

Davidson Current 30
Deep Rover 6, 20, 36, 43
deep seafloor 25, 58, 60
Deep Submergence Laboratory 12
deep-sea biodiversity 73
deep-sea biology 20, 32
deep-sea ecology 37
deep-sea exploration 6-21
Deepstaria enigmatica 57
dragonfish 17, 40-41

echolocation 47
echo-sounders 24
eel 42, 57
eel, gulper 57
El Niño 11, 32
elephant seals 56
Enypniastes eximia 68
epifauna 63
euphausiid 36, 38, 50, 67

fangtooth fish 10
filter-feeders 36, 63, 67
fisheries, deep-sea 75
fishes, deep-sea 72, 75

Galapagos 25
Galapagos rift zone 65
Galathea 15
galatheid crab 70
Galiteuthis phyllura 41
gas-filled swim bladder 71
geology 12, 22
Gnathophausia ingens 54, 57
Godzilla 65
great white shark 56
Gulf of the Farallones 74

hake 42, 43
Harbor Branch Oceanographic Institution (HBOI) 13
hatchetfish, silver 38
HMS Challenger 5, 14, 15, 24, 43, 57, 66
holothurians 68
hot vents 12, 13, 65
Houot, Georges 17
hydrothermal 28
hydrothermal vents 60, 65

Idiacanthus antrostomus 40
IFREMER 13
Indian Ocean 17

Japan Marine Science and Technology Center (JAMSTEC) 13
Jason 12
Johnson-Sea-Link I & II 13
Juan de Fuca Ridge 65

Kaiko 13, 25
Kiyohimea 45, 51, 52
krill 8, 10, 36, 40, 49, 67

lanternfish 38, 40
larvaceans 8, 36, 36, 37
lobsters, squat 70
Loligo opalescens 9
low-oxygen levels 10
luminescence 9, 19, 54

Macrouridae 71
Mariana Trench 18, 25
marine snow 8
Matsumoto, George 45
Mediterranean 17
medusa, medusae 31, 49, 52, 64
Megalodicopia hians 67
Melanostomiatidae 17
Merluccius productus 42
Mid-Atlantic Ridge 13, 22
mid-ocean ridge system 13
midwater 8, 12, 20, 34, 53
Mir, I & II 11, 13, 18
Mirounga angustirostris 47
Monterey Bay 11, 12, 20, 25, 28, 32, 34, 37, 40, 43, 45, 49, 50, 52, 54, ,57, 58, 60, 63, 73, 75
Monterey Bay Aquarium 11, 31
Monterey Bay Aquarium Research Institute (MBARI) 11, 12, 20, 31, 45, 57, 60
Monterey Canyon 11, 12, 20, 31, 70
Monterey submarine canyon 26
Moss Landing Harbor 26
Munida quadrispina 70
Myctophidae 38
mysids 10, 54

Nanomia bijuga 50, 51
Nautile 13
nematocysts 52
Nemichthys scolopaceus 41
New York Zoological Society 16
Nezumia 73
Northern elephant seal 47, 56

Oikopleura villafrancae
ontogenetic vertical migration 54
ophiuroids 69
Orcinus orca 56
owlfish 9, 57
oxygen-minimum layer 28, 49, 54

Pacific dragonfish 40
Packard, David 11, 45
Paralepididae 18
Parmaturus xaniurus 63
particle feeders 38, 44
pharmacology 73
photophores 38, 40
photosynthesis 22, 60, 75
Physeter catodon 47
phytoplankton 22, 28, 30, 32, 34, 36, 50, 53, 58, 75
phytoplankton grazers 30, 38, 44
Piccard, Auguste 17
Piccard, Jacques 18, 25
planktonic plants 22
Poeobius meseres 8, 9
Point Lobos 11
pollution 75
Praya dubia 6, 8, 49, 50
pressure 22
pycnogonid sea spider 69
Pycnogonidia helianthoides 69

rattails 72, 73
rays 71
remotely operated vehicles (ROVs) 11,12, 13, 20, 25, 31, 45, 52
ridges 12

Saccopharynx lavenbergi 57
satellite-borne radar 24
Scripps Institution of Oceanography 12
sea cucumbers 68, 73
sea fans 64, 66
sea pens 64, 66, 69
sea stars 14, 64
seafloor 24, 68, 69
seep communities 11
seep flow 60, 40
Sergestes similis 40, 41
sergestid shrimp 40, 41
sharks 71
Shinkai 6500 13
shrimp 63
skate 71
snipe eel 41
Solmissus 52
Soquel Canyon 27
South Pacific
Southampton Oceanography Center 13
sponges 59, 63, 64, 69
squid, cockatoo 41
squid, market 9
squid, vampire 57
squid 5, 9, 10
squid, giant 5, 47
siphonophore 6, 48, 49, 50, 51
salps 9, 36, 37, 48
submersible 6, 8, 9, 10, 13, 53, 57, 65, 74
submarine canyon 20, 22, 26, 58, 63, 67

tectonics 25
thornyhead 75
Tiburon 11, 12
trenches 25
Trieste 18, 25, 68
tripod fish 17, 73
tunicates 67
turbidity 11, 28, 30

undersea vehicles 17, 42, 53, 70
University of California at Santa Barbara 64
University of Washington 12
upwelling 28, 30, 32

Vampyroteuthis infernalis 54, 57
Ventana 11, 12, 20, 45, 51, 60, 67, 70
volcanoes 12

Walsh, Don 18, 25
Western Flyer 21
whales 47, 49
Wildlife Conservation Society 19
Woods Hole Oceanographic Institution (WHOI) 12, 13
worms 14, 61, 65